SpringerBriefs in Statistics

More information about this series at http://www.springer.com/series/8921

Robert J. Mislevy · Geneva Haertel
Michelle Riconscente · Daisy Wise Rutstein
Cindy Ziker

Assessing Model-Based Reasoning using Evidence-Centered Design

A Suite of Research-Based Design Patterns

 Springer

Robert J. Mislevy
Educational Testing Service
Princeton, NJ
USA

Geneva Haertel
SRI International
Menlo Park, CA
USA

Michelle Riconscente
Designs for Learning, Inc.
San Francisco, CA
USA

Daisy Wise Rutstein
SRI International
Menlo Park, CA
USA

Cindy Ziker
SRI International
Menlo Park, CA
USA

ISSN 2191-544X   ISSN 2191-5458   (electronic)
SpringerBriefs in Statistics
ISBN 978-3-319-52245-6   ISBN 978-3-319-52246-3   (eBook)
DOI 10.1007/978-3-319-52246-3

Library of Congress Control Number: 2017931568

Printed on acid-free paper

This Springer imprint is published by Springer Nature
The registered company is Springer International Publishing AG
The registered company address is: Gewerbestrasse 11, 6330 Cham, Switzerland

To Geneva: A keen researcher, a delightful collaborator, a dear friend (1947–2016)

Preface

Model-based reasoning is essential to navigating the complexities of the real world. Though the literature on model-based reasoning has been available for many years, the advent of the Next Generation Science Standards (NGSS) has increased interest in how students can effectively develop model-based reasoning abilities. The vision of these standards originated with "The Science Framework for K-12 Education" (NRC Framework, 2012), which advocates for the deep integration of the practices of science with the understanding of core ideas and crosscutting concepts (NRC, 2014). The Framework identifies eight practices of science and engineering that are essential for all students:[1]

1. Asking questions (for science) and defining problems (for engineering);
2. Developing and using models;
3. Planning and carrying out investigations;
4. Analyzing and interpreting data;
5. Using mathematics and computational thinking;
6. Constructing explanations (for science) and designing solutions (for engineering)
7. Engaging in argument from evidence; and
8. Obtaining, evaluating, and communicating information.

Developing assessments that target these practices will require tasks that elicit evidence about how students integrate their knowledge of disciplinary core ideas, apply scientific practices, and build connections across ideas (NRC, 2012). In this manuscript, we focus on model-based reasoning as related to the NGSS practice of developing and using models. The National Science Teachers' Association (NSTA) describes developing and using models as "a practice of both science and engineering ... to use and construct models as helpful tools for representing ideas and explanations. These tools include diagrams, drawings, physical replicas,

---

[1]*Next Generation Science Standards: For States, By States* (2013) Chapter: Appendix F: Science and Engineering Practices in the Next Generation Science Standards

mathematical representations, analogies, and computer simulations" (NSTA, 2016). The practices of defining problems and designing solutions for engineering present new opportunities to engage learners in aspects of model-based reasoning, in both instruction and assessment.

In these sections, we describe an approach to designing assessments of model-based reasoning that draws on recent developments in several areas. The first is research on science learning, and, in particular, learning to reason with and through models. A second is cognitive psychology, which highlights both cognitive and social aspects of how people, as individuals and as communities develop models and use them to solve problems and extend their knowledge. The third area is developments in technology, which enable us to build interactive simulations for students, for both learning and assessment, and to make sense of the rich data that can be captured.

The fourth area, the center of this brief, is advances in assessment theory. The first three developments mentioned above expand our knowledge base, widen our vision of what assessment can be, and give us technologies to create richer and more valid assessments. The challenge is how to effectively realize the potential of these advances in practice.

Educational assessment is itself experiencing a renaissance of sorts. Although technological developments have provided improved task environments and psychometric methods, the key development has been to recognize assessment not as a simple exercise in measurement, but as the construction of an argument: What do the particular, situated, noisy observations in a task tell us about students' understandings? About the knowledge structures and activity patterns they can marshal to address what kinds of situations? How does their performance depend on their previous experiences, and how can we sort out the meanings of complex performances in complex tasks? How can we manage task design when multiple aspects of knowledge and skill are involved, and when they may interact differently with different students? How can we design tasks that integrate the broad array of new insights about learning, about the nature of science, and about the technologies that are available for new forms of assessment?

The framework of this brief is a particular approach to these questions, called evidence-centered assessment design. We describe and illustrate a support tool for task development called an assessment design pattern. The suite of design patterns presented here integrate design issues and design choices for seven aspects of building and working with models in science and engineering: Model Formation, Model Use, Model Elaboration, Model Articulation, Model Evaluation, Model Revision, and Model-Based Inquiry. They can be used as stand-alone patterns, or in combinations for in-depth investigations. The key is to organize them as assessment arguments, to enable clear and coherent reasoning about what students know and can do. Assessment arguments provide much-needed structure for coordinating the moving parts of models, cognition, performances, and technology to render integrated and valid insights into student learning. These are tools to help us do it better.

Our goal for this work is to generate a reference that provides insights on design decisions that must be addressed to develop assessments of model based reasoning. We hope these *de-sign patterns* and the framework they are created in can support the design of model-based reasoning in NGSS-inspired assessments, and assessments to come as long as model-based reasoning remains integral to science.

Princeton, USA   Robert J. Mislevy
Menlo Park, USA   Geneva Haertel
Redwood City, USA   Michelle Riconscente
Menlo Park, USA   Daisy Wise Rutstein
Menlo Park, USA   Cindy Ziker

Acknowledgements

Design patterns were introduced in a series of collaborative projects collectively called Principled Assessment Design for Inquiry, or PADI, with SRI International as prime contractor and Geneva Haertel as Principal Investigator. This volume draws on research supported by (1) the Interagency Educational Research Initiative (IERI) under grants REC-0089122, Principled Assessment Design for Inquiry Planning Grant, and REC-0129331, PADI Implementation Grant; (2) the National Science Foundation under grant DRL-0733172, Application of Evidence-Centered Design to State Large-Scale Science Assessment; and (3) the under Grant No. R324A070035, Principled Science Assessment Design for Students with Disabilities. Any opinions, findings, and conclusions or recommendations expressed in this material are those of the authors and do not necessarily reflect the views of the National Science Foundation or the U.S. Department of Education.

Material in this publication is based upon work supported by the United States Government under Contract No. W911NF-14-2-0062. Any opinions, findings and conclusions or recommendations expressed in this material are those of the authors and do not necessarily reflect the views of the U.S. Government. The authors would like to thank the following for their contributions to the works described in this brief: Alex Kernbaum, Karen Koenig, Maria-Isabel Carnasciali, Srujal Patel, Ian Chen, Nate Koenig and the Open Source Robotics Foundation.

Contents

About the Authors

Robert J. Mislevy holds the Frederic M. Lord Chair in Measurement and Statistics at Educational Testing Service, and is Professor Emeritus of Measurement, Statistics, and Evaluation at the University of Maryland. He earned his Ph.D. at the University of Chicago in 1981. His research applies developments in statistics, technology, and cognitive science to practical problems in educational assessment. His work includes a multiple-imputation approach for integrating sampling and test-theory models in the National Assessment of Educational Progress (NAEP), an evidence-centered framework for assessment design, and simulation- and game-based assessments, in collaboration with GlassLAB, the Cisco Networking Academy, and ETS colleagues.

Dr. Mislevy has received NCME's Award for Career Contributions, AERA's E.F. Lindquist Award for contributions to educational assessment, the International Language Testing Association's Messick Lecture Award, AERA Division D's Robert L. Linn Distinguished Address Award, and the NCME's Award for Technical Contributions to Educational Measurement three times. He is a member of the National Academy of Education and a past president of the Psychometric Society. His publications include *Bayesian Psychometric Methods* (Chapman & Hall/CRC, 2016) with Roy Levy, *Bayesian Networks in Educational Assessment* (Springer, 2015) with Russell Almond, Linda Steinberg, Duanli Yan, and David Williamson, and the chapter on cognitive psychology in the fourth edition of *Educational Measurement*.

Geneva Haertel, a pioneer in the application of Evidence-Centered Design (ECD) in large-scale assessment and the practical application of using assessments to improve student learning, wrote publications that focus on the development of valid and reliable assessments for use with a wide range of students, including students with special needs. She published over 60 articles and book chapters on assessment design, student learning, and conditions promoting student achievement. Her research explored the inferential validity, reliability and effectiveness of formative assessments embedded within games. From 2003 to 2008, Dr. Haertel was the SRI Principal Investigator of a series of projects funded by the National Science Foundation, the Institute of Education Sciences, and the Gates and

MacArthur Foundations, that resulted in the development of the online assessment design system referred to as PADI (Principled Assessment Designs for Inquiry).

Dr. Haertel earned her Ph.D. in Education/Educational Psychology at Kent State University and a Bachelor's of Science degree in Education from Kent State University. As the Director of Assessment Research and Design at the Center for Technology in Learning, Dr. Haertel was named a Fellow at SRI International in 2015.

Michelle Riconscente is President of Designs for Learning, a consultancy firm at the intersection of learning, assessment, and digital interactive design. An expert in evidence-centered design, she leads and advises innovative projects that run the spectrum from interactive learning design and embedded assessments, to independent research and evaluation. Previously, as Director of Learning and Assessment at GlassLab, she provided strategic leadership that resulted in valid and reliable game-based assessments of complex competencies such as proportional reasoning, problem solving, and argumentation. The author of over 100 publications and presentations, her clients include Michelson Runway, Digital Promise, Harvard University, MIT Education Arcade, Motion Math, UCLA's CRESST, and Teaching Channel. She holds a Ph.D. in Educational Psychology from the University of Maryland, College Park and a bachelor's degree in Mathematics-Physics from Brown University.

Daisy Wise Rutstein is Education Researcher in SRI International's Center for Technology in Learning. Dr. Rutstein's work focuses on the application of Evidence-Centered Design (ECD) to develop assessments. She moves through the different stages of the development process, starting with the initial conception of the assessment and moving through the development of items, the creation of a complete assessment, and the validation of the assessment. Dr. Rutstein's work in this area has included the development of design patterns, scenarios, and items as well as the identification and application of measurement models for these tasks.

Dr. Rutstein is the lead developer of assessments on several projects in computer science, mathematics, and science including the lead developer for Computational Thinking for the ECS curriculum under the PACT grants (NSF awards CNS-1433065 and DRL-1418149). Dr. Rutstein received her Ph.D. in measurement, statistics, and evaluation from the University of Maryland.

Cindy Ziker is Senior Researcher in the Center for Technology in Learning at SRI International. She earned her Ph.D. in Educational Psychology at Arizona State University and a Graduate Certificate in Large-Scale Assessment from the University of Maryland. Her research focuses on the application of evidence-centered design to the development of technology-enhanced performance tasks and the use of simulation as applied to assessment. She leads validity and evaluation studies of formative and summative assessment systems for state departments of education and large urban school districts. From 2003 to 2006, Dr. Ziker served on a Technical Work Group that evaluated the National Assessment of Educational Progress for Congress. Her publications include "An Introduction to the Evaluation of NAEP", with Suzanne Lane, Bruno D. Zumbo, Jamal Abedi, Jeri

Benson, John Dossey, Stephen N. Elliot, Michael Kane, Robert Linn, Michel Rodriguez, Greg Schraw, Jean Slattery, Veronica Thomas, and Joe Willhoft (2009), and a chapter in *Meeting the Challenges to Measurement in an Era of Accountability* with Geneva D. Haertel, Terry P. Vendlinski, Daisy Rutstein, Angela DeBarger, Britte H. Cheng, Christopher J. Harris, Cynthia D'Angelo, Eric B. Snow, Marie Bienkowski, and Liliana Ructtinger (Routledge, 2016).

Chapter 1
Introduction

Abstract Understanding, exploring, and interacting with the world through models characterizes science in all its branches and at all levels of education. Model-based reasoning is central to science education and thus science assessment. Building on research in assessment, science education, and learning sciences, we present a set of design patterns to help assessment designers, researchers, and teachers create tasks for assessing aspects of model-based reasoning. This chapter provides a rationale for the design patterns and evidence-centered assessment design, lays out the structure of the book, and introduces two running examples of inquiry assessments that will be used to illustrate the ideas.

Models are fundamental to science. The centrality of Newton's laws, the double helix model of DNA, and the Lotka-Volterra model of predator-prey interaction are cases in point (Frigg & Hartmann, 2006). But it is not simply the knowledge contained in models that matters; even more important are the ways models are used to carry out scientific practices. Scientists and engineers build, test, compare, and revise models. They use models to organize experience, guide inquiry, communicate with one another, and solve practical problems (Lehrer & Schauble, 2006).

The National Science Education Standards (NSES; National Research Council, 1996) highlight "Evidence, Models, and Explanation" as a unifying theme for science education, spanning grade levels and science domains. Similarly, the Next Generation Science Standards (NGSS; National Research Council, 2012; NGSS Lead States, 2013a, b) incorporate dimensions of content, practices and cross-cutting themes. The concepts of models and model-based reasoning appear across both the individual practices (developing and using models) and the cross-cutting practices (including system models). To guide and evaluate students' progress with respect to these standards, it is therefore important to be able to assess their proficiencies in reasoning with and about models.

While it is fairly straightforward to assess students' familiarity with concepts, terminology, and calculation, assessing model-based reasoning is more challenging (National Research Council, 2001; Pellegrino, 2013). How can we devise occasions and settings for students to display their capabilities to build, critique, revise, and

© The Author(s) 2017
R.J. Mislevy et al., *Assessing Model-Based Reasoning using Evidence-Centered Design*, SpringerBriefs in Statistics, DOI 10.1007/978-3-319-52246-3_1

use models, to understand, explain, predict, and produce effects in the natural world? How might we evaluate the cycles of observing, hypothesizing, and reformulating that characterize inquiry using models? Are there principles and approaches to help us assess model-based reasoning across the diversity of models used in different branches of science and across levels of education from the primary grades to postsecondary study?

Many of these challenges can be addressed with the help of *design patterns*. *Design patterns* are used in architecture and software engineering to characterize recurring problems and approaches for solving them, such as Workplace Enclosure for house plans (Alexander, Ishikawa, & Silverstein, 1977) and Interpreter for object-oriented programming (Gamma, Helm, Johnson, & Vlissides, 1994). This brief provides a suite of assessment *design patterns* to support the design of tasks to assess model-based reasoning.

These *design patterns* help domain experts and assessment specialists "fill in the slots" of an assessment argument built around recurring themes in learning (Liu & Haertel, 2011). The particular form of *design patterns* presented here were developed in the Principled Assessment Design for Inquiry (PADI) project (Mislevy et al., 2003, b). Table 1.1 lists technical reports developed in the PADI projects that present a variety of design patterns and related supports.

In addition, DeBarger, Krajcik, Harris, and Penuel (2013) and Harris, Krajcik, Pellegrino, and McElhaney (2016) show how to use design patterns to develop tasks that jointly address disciplinary core ideas, crosscutting concepts that span science domains, and science practices, as advocated in the NGSS.

Chapter 2 is an overview of model-based reasoning. It draws on studies of model-based reasoning in science, including Gobert and Buckley (2000), Grosslight, Unger, Jay, and Smith (1991), Schwarz et al., (2009), Snir, Smith, and Raz (2003), Spitulnik, Krajcik, and Soloway (1999), and Stewart and Hafner (1994). It highlights interaction and iteration in the ways scientists and engineers use models—continually constructing and reconstructing correspondences between general structures and unique real-world situations.

Chapter 3 describes the "evidence-centered" approach to assessment under which *design patterns* are conceived, followed by the attributes of a PADI *design pattern*.[1] The remainder of the brief presents the suite of *design patterns*. Table 1.2 lists the design patterns with brief descriptions. They are presented in tabular form in Appendix 1. Following an overview in Chap. 4, subsequent sections address each aspect of model-based reasoning. Chapters 5 through Chap. 11 discuss, in turn, Model Formation, Model Use, Model Elaboration, Model Articulation, Model Evaluation, Model Revision, and Model-Based Inquiry. They provide additional discussion and illustrations of the summary forms in the appendix. These design

---

[1]Fuller discussions of ECD appear in Mislevy, Steinberg, and Almond (2003), Mislevy and Riconscente (2006), and technical reports from the Principled Assessment Design for Inquiry (PADI) project (http://padi.sri.com/publications.html). For application of ECD to simulation-and game-based assessments, see Mislevy (2013), Mislevy et al. (2014), and Riconscente, Mislevy, & Corrigan (2015).

Table 1.1 Technical reports developed by PADI projects

Baxter and Mislevy (2005). *The case for an integrated design framework for assessing science inquiry (PADI Technical Report 5)*. Menlo Park, CA: SRI International

Brecht, Cheng, Mislevy, Haertel, and Haynie (2009). *The PADI System as a Complex of Epistemic Forms and Games (PADI Technical Report 21)*. Menlo Park, CA: SRI International

Cheng, Ructtinger, Fujii, and Mislevy (2010). *Assessing Systems Thinking and Complexity in Science (Large-Scale Assessment Technical Report 7)*. Menlo Park, CA: SRI International

Colker, Liu, Mislevy, Haertel, Fried, and Zalles (2010). *A Design Pattern for Experimental Investigation (Large-Scale Assessment Technical Report 8)*. Menlo Park, CA: SRI International

DeBarger and Riconscente (2005). *An example-based exploration of design patterns in measurement (PADI Technical Report 8)*. Menlo Park, CA: SRI International

DeBarger and Snow (2010). *Design Pattern on Model Use in Interdependence among Living Systems (Large-Scale Assessment Technical Report 13)*. Menlo Park, CA: SRI International

Fulkerson, Nichols, Haynie, and Mislevy, (2009). *Narrative Structures in the Development of Scenario-Based Science Assessments (Large-Scale Assessment Technical Report 3)*. Menlo Park, CA: SRI International

Gotwals and Songer (2006). *Cognitive Predictions: BioKIDS Implementation of the PADI Assessment System (PADI Technical Report 10)*. Menlo Park, CA: SRI International

Haertel, Haydel DeBarger, Villalba, Hamel, and Mitman Colker (2010). *Integration of Evidence-Centered Design and Universal Design Principles Using PADI, an Online Assessment Design System (Assessment for Students with Disabilities Technical Report 3)*. Menlo Park, CA: SRI International

Haynie, Haertel, Lash, Quellmalz, and DeBarger (2006). *Reverse Engineering the NAEP Floating Pencil Task Using the PADI Design System (PADI Technical Report 16)*. Menlo Park, CA: SRI International

Liu and Haertel (2011). *Design Patterns: A Tool to Support Assessment Task Authoring (Large-Scale Assessment Technical Report 11)*. Menlo Park, CA: SRI International

Mislevy and Haertel (2006). *Implications of Evidence-Centered Design for Educational Testing (PADI Technical Report 17)*. Menlo Park, CA: SRI International

Mislevy, Haertel, Cheng, Rutstein, Vendlinski, Murray, et al. (2013). *Conditional Inferences Related to Focal and Additional Knowledge, Skills, and Abilities (Assessment for Students with Disabilities Technical Report 5)*. Menlo Park, CA: SRI International

Mislevy, Hamel, Fried, Gaffney, Haertel, Hafter, et al. (2003). *Design patterns for assessing science inquiry (PADI Technical Report 1)*. Menlo Park, CA: SRI International

Mislevy, Liu, Cho, Fulkerson, Nichols, Zalles, et al. (2009). *A Design Pattern for Observational Investigation Assessment Tasks (Large-Scale Assessment Technical Report 2)*. Menlo Park, CA: SRI International

Mislevy and Rahman (2009). *Design Pattern for Assessing Cause and Effect Reasoning in Reading Comprehension (PADI Technical Report 20)*. Menlo Park, CA: SRI International

Nichols and Fulkerson (2010). *Informing Design Patterns Using Research on Item Writing Expertise (Large-Scale Assessment Technical Report 9)*. Menlo Park, CA: SRI International

Seeratan and Mislevy (2009). *Design Patterns for Assessing Internal Knowledge Representations (PADI Technical Report 22)*. Menlo Park, CA: SRI International

Snow, Fulkerson, Feng, Nichols, Mislevy, and Haerte (2010). *Leveraging Evidence-Centered Design in Large-Scale Test Development (Large-Scale Assessment Technical Report 4)*. Menlo Park, CA: SRI International

(continued)

Table 1.1 (continued)

Zalles, Haertel, and Mislevy (2010). *Using Evidence-Centered Design to Support Assessment, Design and Validation of Learning Progressions (Large-Scale Assessment Technical Report 10).* Menlo Park, CA: SRI International

Zhang, Mislevy, Haertel, Javitz, Murray, and Gravel, (2010). *A Design Pattern for a Spelling Assessment for Students with Disabilities (Assessment for Students with Disabilities Technical Report 2).* Menlo Park, CA: SRI International

Table 1.2 Aspects of Model-Based Reasoning in Science

Aspect	Definition
Model formation	Establishing a correspondence between some real-world phenomenon and a model, or abstracted structure, in terms of entities, relationships, processes, behaviors, etc. Includes determination of the scope and grain-size to model, which aspects of the situation(s) to address and which to leave out
Model use	Reasoning through the structure of a model to make explanations, predictions, conjectures, etc.
Model elaboration	Combining, extending, and adding detail to a model. Establishing correspondences across overlapping models into larger assemblages. Fleshing out more general models with more detailed models
Model articulation	Connecting meaning of physical or abstract systems across multiple representations. Representations may take qualitative or quantitative forms. Notably relevant in models with quantitative and symbolic components, such as the conceptual and mathematical aspects of physics models
Model evaluation	Assessing the correspondence between the model components and their real-world counterparts with emphasis on anomalies and important features not accounted for in the model
Model revision	Modifying or elaborating a model for a phenomenon in order to establish a better correspondence. Often initiated by model evaluation procedures
Model-based inquiry	Working interactively between phenomena and models, using all aspects of the above. Emphasis on monitoring and taking actions with regard to model-based inferences vis-à-vis real-world feedback

patterns can be used separately to develop tasks that target particular aspects of model-based reasoning, or in concert to develop more complex multi-stage or iterative investigations.

To further illustrate the application of this approach, discussions of two iterative investigation tasks appear at several points in the presentation. They show how *design patterns* can contribute in combinations to develop complex tasks and evaluate students' performances. The first of these is based on a genetics investigation in a curriculum devised by Stewart and his colleagues (Johnson & Stewart, 2002). The introduction to this genetics investigation is given in Box Genetics-1. The second running example is a design task from a massive online open course (MOOC). It illustrates the use of model-based reasoning in engineering. It is introduced in Box Robotics-1.

More focused tasks specific to particular aspects of model-based reasoning appear in the discussions of the various *design patterns*. A variety of models, content domains, task types, and educational levels are included to suggest the breadth of applicability of *design patterns* to support task design.

Chapter 12 summarizes the rationale for using *design patterns* to help develop assessments of model-based reasoning. We note their relevance to standards-based assessment, instruction, and large-scale accountability testing.

Box Genetics-1. Introduction

Stewart and Hafner developed a course containing laboratories for the study of baseline genetics models. These laboratories included the use of Jungck and Calley's *Genetics Construction Kit* (GCK 1985), a software simulation program that includes the ability to construct customized problems to study different genetics phenomena (Stewart, Hafner, Johnson, & Finkel, 1992).

At the start of the course, students learned about the **development of models** and read an abridged version of Mendel's paper (Johnson & Stewart 2002). They were then visited by a graduate student dressed as Mendel who taught them about the simple dominance model. Students were subsequently tasked with using the GCK to determine whether the simple dominance model could be applied to test crossing organisms. Using a box containing several specimens with a given trait, students performed a cross and examined the resulting output to identify which crosses produced which traits. They then responded to questions regarding their findings, such as which trait appeared to be the most dominant. The figure below, which represents an early stage in problem-solving, is taken from the Virtual Genetics Kit, software based on the GCK (http://intro.bio.umb.edu/VGL/index.htm).

The initial task represented a case of **model use**, in which students applied a known model to a given set of data. The next task presented data that did not strictly adhere to the simple dominance model in order to advance students' understanding of model-based reasoning. Students were prompted to **evaluate** the fit of the data to the known model and, upon discovering that the known model was inadequate, to **revise** the existing model to account for the observed deviations. Working in groups, students conducted their analyses and developed revised models they tested using crosses provided by the other groups. Each group then presented their solution. This process was repeated for different models, such as the codominance and the multiple allele models. Data were collected from each round and used to assess each student's proficiency in model revision. These data included recordings of the research group interactions both internally and with the instructor; lab books; interactions with the software such as the sequence of actions performed; and a written description of the group's final model. Later sections of this paper provide details on additional examples and associated assessments of student ability.

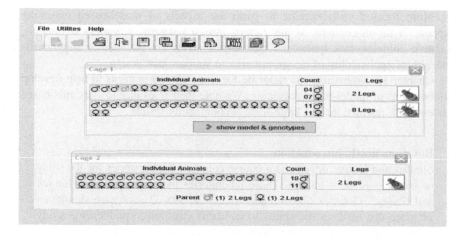

Box Robotics-1. Introduction

The SiMPLE project (IISES 2016; DARPA 2016) created graphical tools for assembling virtual simulation models and changing their parameters to iteratively test their performance, using Gazebo simulation software that contains virtual models of machines, structures, and environments. Gazebo enables students to visualize the behavior of physical models while experimenting with the dynamic and interactive parameters of a design. The example is drawn from an exercise embedded in a Massive Open Online Course (MOOC) that provides simulation tools, instructional material, and online support capabilities that include informal, formative assessment of students' activities, including reasoning with simulated and physical models.

A key feature of the SiMPLE course materials is the use of multiple representations to accelerate learning (Fig. R1). These representations include: a 3D world view to enable visualization of model dynamics and interactions within the simulated world environment, a schematic view that allows for easy comparisons between disparate systems, a model editor view that shows the kinematics of the model, and a physical representation that is created using a robot kit. A graphing utility tool provides visual representations to enhance learner diagnosis of design flaws by plotting simulation properties over time; this tool allows users to quickly optimize simulations and make quantitative comparisons (see Fig. 1.1). Students can use this tool to explore how components would function, test their designs, and modify complex systems.

The task we address involves model-based reasoning in engineering: Students build a motorized rover, for challenges such as hill-climbing and tug-of-war. They first **create**, **evaluate**, and **revise** schematic models of the device in the simulation space, then use components to **develop**, **evaluate**,

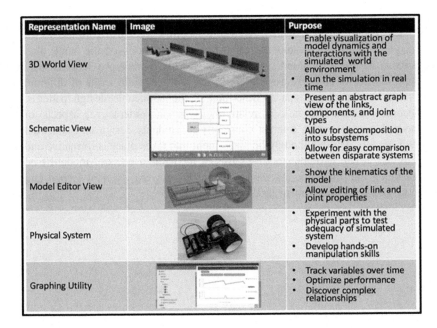

Representation Name	Image	Purpose
3D World View		• Enable visualization of model dynamics and interactions with the simulated world environment • Run the simulation in real time
Schematic View		• Present an abstract graph view of the links, components, and joint types • Allow for decomposition into subsystems • Allow for easy comparison between disparate systems
Model Editor View		• Show the kinematics of the model • Allow editing of link and joint properties
Physical System		• Experiment with the physical parts to test adequacy of simulated system • Develop hands-on manipulation skills
Graphing Utility		• Track variables over time • Optimize performance • Discover complex relationships

Fig. R1 SiMPLE leverages multiple representations to enhance learning

Fig. R2 Physical gear box and motor

and **revise** a physical device based on what they have learned with their models in the simulation space. Although the task also involves designing the electrical circuit for the battery-powered motor, we will focus on modeling the gear box and weight distribution in the hill-climbing challenge: designing and building a rover, then driving it up a ramp. Students simulate each of the gear ratios to see the results, and then follow the steps needed to build a physical gear box that includes a motor (Fig. R2).

This robotics task has three interesting features for illustrating the use of design patterns. First, students carry out investigations of the same phenomena in the simulation and physical worlds, which provides us with opportunities to compare assessment in the two realms in terms of different choices in the design spaces described in the applicable design patterns. Second, because the task is relatively open and interactive, aspects of model-based reasoning interact constantly as students design and construct their models. Third, although the assessment that takes place is formative and informal, it can be understood in terms of the same aspects of model-based reasoning and the same assessment design framework as more formal and more structured assessment tasks.

Chapter 2
Model-Based Reasoning

Abstract Model-based reasoning consists of cycles of proposing, instantiating, checking, revising to find an apt model for a given purpose in a given situation, and reasoning about the situation through the model. Results from cognitive research can help us understand and assess both the experiential and reflective aspects of model-based reasoning. This chapter reviews research on model-based reasoning and the inquiry cycle to define aspects of model-based reasoning that can be used to guide assessment design.

Broadly speaking, inquiry is the process by which scientists and engineers formulate and investigate questions about the natural world in order to formulate answers, explanations, predictions, designs, or theories (NSES). Developing inquiry skills means being able to reason through fundamental concepts and relationships to understand and interact with particular real-world situations—in short, reasoning through models. Because scientific models embody hard-won, powerful, knowledge about how the world works, students do need to learn about important models in disciplinary areas of science. Many science-education researchers regard model-based reasoning as a pivotal way to unify content, the activities of inquiry, and teaching and learning (Stewart and Gobert, op cit.; Buckley, 2012); Clement 2000); Gilbert & Justi, 2016; Hestenes, 1987).

2.1 Scientific Models

A model is a simplified representation that focuses on certain aspects of a system (Ingham & Gilbert, 1991). Its entities, relationships, and processes constitute its fundamental structure. They provide a framework for reasoning across any number of unique real-world situations. The model abstracts salient aspects of the situations and goes further by viewing them as instances of recurring mechanisms, causal relationships, or connections across scales or time points that are not apparent on the surface. It formalizes experience usually by many people, tested, argued,

© The Author(s) 2017
R.J. Mislevy et al., *Assessing Model-Based Reasoning using Evidence-Centered Design*, SpringerBriefs in Statistics, DOI 10.1007/978-3-319-52246-3_2

extended, and accumulated sometimes over centuries. Frigg and Hartmann (2006) provide an overview of models in science, and Harrison and Treagust (2000) give a typology of models as they are used in STEM education and in practice.

This brief concerns the explicit models that scientists create and use, and are targets of learning in science and engineering. The focus is not simply models as representations, but models as epistemic tools: Ways to understand the world, to interact with it, and to change it (Gilbert & Justi, 2016). A scientific model is a community resource—a particularly technical special case of what cognitive anthropologists call cultural models (Strauss & Quinn, 1998). The system of concepts, relationships, and processes that constitute a model extends beyond the mind of any individual. It is manifest in books and tools, in activities both formal and informal, in ways of seeing the world, and in patterns one can interact with the world and others. A web of interrelated ideas and activities spans individuals, is contributed to by many, is used by many more, and is enriched with every use (Latour, 1987). Science education aims to bring students into the community—to acquaint them with key concepts and relationships of important models, to be sure, but further to empower them to interact with the ideas and with people in the practically useful ways that are mediated through scientific models.

A broad conception of models highlights similarities in the kinds of thinking and activities that occur across a range of models. We want to ground *design patterns* on broad similarities in order to support task design across a broad range of content and contexts. More focused *design patterns* can be constructed for particular classes of models and representations. They would provide more focused support for particular areas of science or kinds of tasks.

For our purposes, models can be as simple as the change, combine, and compare schemas in elementary arithmetic (Riley, Greeno, & Heller, 1983), or as complex as quantum mechanics, with multiple forms of representation, advanced mathematical formulations, and interconnections with other physical models. Models can contain or overlap with other models. Relationships among the entities can be qualitative, hierarchical, dynamic, and spatial. Some models concern processes, such as the stages of cell division in meiosis. Some relationships can be qualitative (if Gear A rotates clockwise, Gear B must rotate counterclockwise), and some support quantitative or symbol-system representations and operations (if Gear A has 75 teeth and Gear B has 25, Gear B will rotate three times as fast as A). There can be different models for the same phenomena. The wave and particle models for light connect in some important aspects (amount of energy) but differ in ways that are useful for different problems (diffraction patterns versus the photoelectric effect).

Figure 2.1 suggests some central properties of model-based reasoning (a simplification of Greeno, 1983). The lower left plane (A) shows phenomena in a particular real-world situation. A mapping is established between this situation and, in the center plane (B), the patterns expressed in terms of the entities, relationships, and properties of the model. This is the "semantic" layer of the model. Reasoning is carried out in these terms. This process constitutes the reconception of the situation shown at the lower right (E). It synthesizes particulars of the situation with the abstracted structure of the model—a "blended space" for reasoning, as Fauconnier

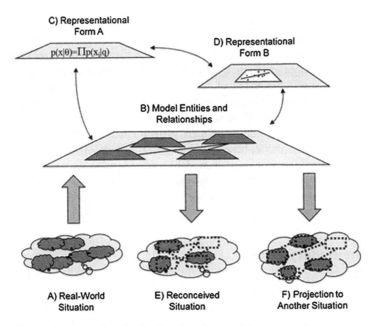

Fig. 2.1 Reconceiving a real-world situation through a model

and Turner (2002) call it. The processes and relationships of the model are used to make inferences such as explanations about the current real-world situation, and inferences about other situations (F) such as predictions or designs for artifacts (Swoyer, 1991). Above the layer of entities and relationships are symbol systems (C and D are two of possibly several) that further support reasoning in the semantic layer of this model, such as diagrams, matrix algebra, and computer programs.

Figure 2.1 also suggests properties that are important for understanding how models are used. The real-world situation is depicted as nebulous, whereas the entities and relationships in the model are crisp and well defined. Not all aspects of the real-world situation have corresponding representations in the model. On the other hand, the model conveys ideas and relationships that the real-world situation does not. The situation as reconceived through the model shows a less-than-perfect match to the model, but it provides a framework for reasoning that the situation itself does not.

The validity of a model does not address a two-way relationship between a model and reality, but a four-way relationship among a model, reality, a user, and a purpose (Suárez, 2004). As the statistician George Box said, "all models are wrong, but some are useful." Being able to construct models that suit both the situation and the purpose at hand is central to model-based reasoning. Reasoning within models' narrative spaces and manipulating their symbol systems are important, but they are not enough.

The strategies, the procedures, and the rules of thumb that enable one to put a model to practical use are the kinds of "epistemic games" (Collins & Ferguson, 1993) students must learn if they are to develop their capabilities for reasoning with models. Students learn to reason in these ways *by* reasoning in these ways—in specific and real problems, in classrooms, in projects, in games, in hobbies. Ideally, support and feedback sharpens their reasoning and makes its generalizable structure explicit. Through these experiences, students begin to build increasingly broad and more generally applicable resources for both working with particular models and for the processes for reasoning with models (Schunn & Anderson, 1999).

Assessment of students' thinking and activities helps instructors guide their learning, and helps curriculum developers generate activities that fully reflect the targeted learning. The model-based reasoning *design patterns* bring out essential, recurring aspects of the processes of model-based reasoning, in ways that connect them to assessment arguments and help educators develop tasks to draw them out, whether focusing on particular aspects or on their interplay in investigations.

2.2   The Inquiry Cycle

In traditional science education, students are presented with models and asked to apply them to problems (Stewart & Hafner, 1991). But model-based reasoning in practice is characterized by the processes of proposing, instantiating, checking, and revising to find an apt model in a given situation. A model-based reconception is often provisional. Hypothesized missing elements can be used to evaluate the quality of the representation, and prompt a user to revise or to abandon a particular model. The hypothesized relationships then guide actions that change real-world situations and lead to further cycles of inquiry, understanding, and action. The depiction of the inquiry cycle in Fig. 2.2 (from White, Shimoda, & Frederiksen, 1999) is useful for highlighting aspects of model-based reasoning as they are used in inquiry and as they can be addressed in assessment.

Students can be presented, or propose on their own, a question that can be addressed by the concepts and principles in a scientific domain, then determine

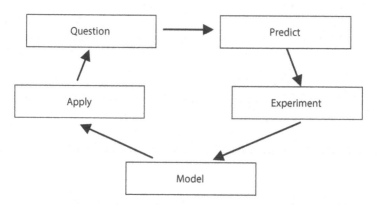

Fig. 2.2 The inquiry cycle

what observations might bear on its solution. They may be presented with, or gather themselves, data about the natural world, then build a model to account for patterns in the data. Once they have formulated a model, they may be asked to test the model by making predictions about further observations and determining whether it holds up in light of new information or requires modifications. If so, the cycle of model-building, model-checking, and model-revision continues, each stage requiring its own particular kind of reasoning.

Typically, students are introduced first to simpler forms of models and inquiry (e.g., provided substantial scaffolding to guide their investigations) and are then gradually exposed to more complex models (as described in the example in Box Genetics-1) and more independent situations for using them (Gotwals & Songer, 2010; Hammer, Elby, Sherr, & Redish, 2005; Redish, 2004; Songer, Kelsey, & Gotwals, 2009).

The multifaceted nature of model-based reasoning holds implications for both instruction and assessment. An instructor's decision to highlight to certain aspects will require assessment attuned to those aspects. The focus of instruction, and thus of assessment, for a new model may initially be reasoning through that model with data that are known to be appropriate. Alternatively, an instructor may want to see students work through cycles of inquiry with a model that is already familiar to the students. These latter tasks allow a focus on the self-monitoring and organizational capabilities required to coordinate the aspects of reasoning that interact when fitting models iteratively.

Students do not develop competence across all aspects of model-based reasoning at the same rate and depth. A student may be more facile with some aspects of inquiry in some content domains than others—and even for different investigations in the same domain (Mislevy, 2017; Ruiz-Primo & Shavelson, 1996). Instructors and assessment designers must consider the interplay between models and model-based reasoning, and where they want to focus attention. For example, an exercise meant to highlight model-checking could use a model familiar to students. An exercise to expand students' capabilities with a new model could employ a model-checking technique that students are familiar with from a previous lesson. The task designer must consider the extent to which declarative knowledge of a model's structure and components—as opposed to reasoning with and through the model—are to be stressed. Making this determination depends not simply on what is in the task but on the relation of that task to the experience of the examinee. This relationship may be known (e.g., as in local assessments embedded in instruction) and leveraged to sharpen the evidentiary focus of a task. Conversely, the relationships may be unknown (e.g., as in large-scale accountability tests), so that information about examinees' substantive knowledge about a model and their capability to use it are confounded. Sect. 2.3.4 says more about how these choices affect the evidentiary value of tasks in different assessment uses.

2.3   Some Relevant Results from Psychology

There are two basic modes of human cognition. Kahnemann (2011) called them "fast thinking" and "slow thinking;" Norman (1993) described them as experiential and reflective: "The experiential mode leads to a state in which we perceive and react to the events around us, efficiently and effortlessly. The reflective mode is that of comparison and contrast, of thought, of decision making. ... Both modes are essential to human performance (p. 15, 20).

Model-based reasoning involves both. As Giere (1987) put it,

> My general view is that scientific theories should be regarded as continuous with the representations studied in the cognitive sciences. There are differences to be sure. Scientific theories are more often described using written words or mathematical symbols than are the mental models of the lay person. But fundamentally they are the same sort of thing (p. 143).

This section notes some results from research in cognitive psychology and learning science that are useful for understanding model-based reasoning, how people become proficient, and then how they might be assessed.

2.3.1   Experiential Aspects of Model-Based Reasoning

A person forming a mental model to understand a situation activates, assembles, and particularizes elements from long-term memory to create an instance of a model that is tailored to the task at hand. Walter Kintsch's "construction-integration" (CI) model of text comprehension (Kintsch, 1998) provides insights into the process. Kintsch and Greeno (1985) apply the CI perspective to understanding reasoning with models. In one of their examples, the models of interest are Change, Combine, and Compare arithmetic schemas, and the problem is figuring out how a problem situation correspond to these models.

For a simple word problem, model formation takes place in working memory, incorporating features of the situation from sensory memory and information from long-term memory. Features of the situation activate elements of long-term memory, which can in turn activate other elements of memory or guide a search for new features in the situation. A person's goals and affective state also influence what models are activated. This construction phase (the C in CI theory) is initiated by features of stimuli in the environment and activates associations from long-term memory-whether or not they are relevant to the current circumstances.

A "situation model" emerges from the integration (the I in CI theory) of mutually reinforcing elements among the immediate stimuli and the retrieved patterns. The situation model constitutes the person's comprehension of the situation. Particular elements of the real-world situation are synthesized with more generalized patterns from that individual's previous experience. Ideally, in the case of scientific models, the person activates appropriate chunks of formal models, and its elements correspond to elements in the real-world situation. Now the situation is comprehended in terms of the salient elements and relationships in the scientific model (Larkin, 1983). This model formation sets the stage for further reasoning by activating, to the extent the person has developed them, associations of many sorts—narratives, representations, procedures, strategies, examples, and personal experiences.

The same cognitive processes also take place when students reason with partial, incomplete, fragmentary, and intuitive building blocks rather than with correct scientific models (diSessa, 1993, calls them phenomenological primitives, or "p-prims"). The resulting situation model again draws on patterns from the student's past experience, which together provide an understanding of the situation upon which to base further reasoning and action. Unlike the situation model of an expert, however, this understanding may be based on superficial features of the situation or misconceptions; for example, the "continuous push" p-prim that an object will keep moving only if some force is continuously applied to it. Such understandings often suffice for everyday life. But they are not cast in terms of coherent conceptions that connect diverse situations and link them to effective procedures and strategies. People reasoning in this way are employing model-based reasoning, but not through the models that are the targets of science instruction.

Successfully forming a cognitive situation model around a scientific model requires not only the availability of the formal elements of the scientific model from long-term memory, but cues to activate them and to then relate them to the real-world situation (Redish, 2004). Experts have more information in long-term memory about models than do novices, but more importantly, they have more effective connections among them—including the conditions of when they are useful (Glaser, Chi, & Farr, 1988). Experts' model formation is streamlined by extensive use, to accommodate more rapid access, larger chunks, and routinized.

For example, Chi, Feltovich, and Glaser (1981) asked novices and experts in physics to sort cards depicting mechanics problems into stacks of similar tasks. Novices grouped problems in terms of surface features such as pulleys and springs. Experts organized their groups around more fundamental principles such as equilibrium and Newton's Third Law, each group containing a variety of spring, pulley, and inclined plane tasks. The experts' categories reflect a well-practiced model formation process for understanding real-world situations in terms of principles that are not apparent on the surface. Their situation models are linked, in turn, to mathematical representations for solving problems (Model Use), for criteria to evaluating its suitability (Model Evaluation), and to strategies and procedures for carrying out these activities.

2.3.2   Reflective Aspects of Model-Based Reasoning

While scientific models can ground an individual's understanding about a situation, they also are cultural tools that people can use to think and act together—a special case of what Wertsch (1998) calls mediated action. Seeing model-based reasoning as action underscores how science is not merely a matter of models, formulas, and procedures, but ways of thinking, talking, and acting in the world, through patterns of knowledge and understanding that have built up within a community of practice.

Processes analogous to the CI model take place in conscious, explicit, model-based reasoning; that is, reasoning among people, using tools and external representations, occurring over minutes, hours, or years rather than milliseconds. Tools and external representations embody key relationships to enable computation and capture intermediate results to help overcome the limitations of working memory (Markman, 1999). The cognitive activation of relevant information in an individual's long-term memory is echoed externally in literature searches and conversations with colleagues. The external counterparts of refocusing a gaze are now generating scatterplots, looking for trends and outliers, and re-expressing residuals in a different format. The elements of a tailored, synthesized, and integrated model can be drawn from different domains, and reconfigured through multiple drafts of an article. The correspondence between the elements of real-world situations and the entities in an instantiated scientific model may require repeated attempts to determine just what to address, at what level of detail, and in what representational form to achieve the goals at hand. These are cycles of Model Formation, Model Evaluation, Model Elaboration, and Model Revision.

Managing one's own activities in their full complexity over time requires being able to reflectively monitor one's progress, evaluate the effectiveness of work, keep track of where one is, and determine next steps. These are metacognitive skills associated with model-based reasoning. White, Shimoda, and Frederiksen (1999) cited Piaget (1976)'s argument that reflecting on one's cognition reflects an advanced stage of development, and Vygotsky's (1978) claim that children progress from relying on others to help regulate their cognition to being able to regulate it themselves. Chapter 11 draws on this work for the *design pattern* for creating tasks to assess how students coordinate aspects of model-based reasoning within more encompassing activities. Self-regulation can be scaffolded as an option to design instruction to help students develop these skill, and to design assessments that either support them or put greater demands on them to assess them at higher levels.

2.3.3   Higher-Level Skills

Educators agree on the importance of higher-order skills such as critical thinking, problem-solving, systems thinking, and, to the present concern, model-based reasoning. There is less agreement on just what these terms mean. What is the nature of such skills, and how are they acquired? How they might be assessed? Research sheds light on the issue, and highlights design decisions that must be made in different ways to make the terms meaningful for particular purposes in particular contexts (the "use cases" described in Sect. 2.3.4).

These results follow from a view of learning as developing resources through experiences in specific contexts (Bransford & Schwartz, 1999; Hammer et al.,

2005). Building resources for, say, model-based reasoning starts in work with particular models, simple ones at first. The work is entwined with knowledge and skills connected with those models, and the particular problems and contexts in the situation at hand. Further experience begins to encompass more complex models, more complicated situations, and more sophisticated reasoning, always in the context of particular models and purposes. To the degree that the more general concepts and representations of working with systems are brought to the surface, learners begin to organize resources that can be adapted more readily to new models and more advanced practices (Schwartz et al., 2009). Students shift from seeing models as correct or incorrect to models as encompassing explanations for multiple aspects of a phenomenon. They develop more nuanced reasons to revise models. More advanced activities present challenges such as constructing a model to aid their own sense-making, and seeing model building as a way to generate new knowledge.

Still, engaging in what would be called "model-based reasoning" in any particular situation will jointly require resources for the substance, the context, and the practices that are involved. It is only through experience with multiple models in multiple contexts that students begin to develop more general capabilities they can bring to bear in new situations (National Research Council, 2000; Perkins & Salomon, 1989).

Constructs like "model formation" and "model revision" thus call out similarities as they appear to an outside observer, across what people do in situations that vary considerably in context and substance. Any assessment of model-based reasoning must therefore always face design decisions about the models, the content, and the context that are at issue. Critical questions for an assessment designer include what students know about the content and context, and what the designer knows about what the students know. *Assessment use cases* are helpful for thinking about these design issues.

2.3.4 Implications for Assessment Use Cases

The term "assessment" spans a broad array of ways and purposes for gathering information about what students know, can do, or might work on next. An assessment use case is a recurring configuration of people, information, contexts, and purposes. Model-based reasoning tasks have an inherent complexity because they necessarily involve some content, some context, and some practices. The interplay among these factors and the relationship to students' backgrounds holds different implications for assessing model-based reasoning in the four use cases described below. Keeping the use case in mind while referring to the design patterns for support helps a designer make appropriate choices. It is not the features of a task alone that determine its evidentiary value, but the match of the task to the purpose and the students who will be assessed (Gorin & Mislevy, 2013).

Use Case 1: Formative assessment during learning activities

In this use case, inferences about students are used for feedback to further learning. It could be to a teacher, a learning system, or the students themselves. A significant factor of a task's value is how it matches up with what is known about students' previous experiences: A task may be quite complex, but for students working with this model at this time, some aspects will be known to be familiar and thus minor sources of challenge. Much of the knowledge that is necessary but irrelevant to the learning target is known to be familiar, and certain aspects of knowledge or modeling processes are targeted as the primary challenge. The evidentiary value of a task under these conditions can be quite high *for the targeted inferences*, because it is matched to local purposes about these students and takes advantage of local knowledge about their current and past experiences.

Use Case 2: Large-scale student-level accountability assessment

Consider a state accountability test where every student in Grade 6 is administered at a randomly-selected set of tasks, to estimate scores for individuals. The tasks are assigned without consideration of the matchups of the previous use case. Research on large-scale performance assessments shows that a student's performance on complex tasks assigned without knowing how the facets of the task match up with the students' previous experiences often does not convey very much information about how she would fare with a different, equally acceptable, task (Linn, 2000). The more diverse the test-takers, the greater the effect. There is low generalizability from how a student performs from one context to another or with one model to another, with respect to what is nominally "the same scientific process skill."

Use Case 3: Summative assessment in a course of instruction

This use case blends features of the two discussed above: assessments are integrated with a course of learning, but are used with higher stakes for individuals, such as a course grade or a certification. The College Board's Advanced Placement (AP) examinations are an example. Like both the accountability tests of Case 2 and the educational surveys of Case 4 below, AP examinations are large-scale assessments, developed and evaluated outside the local learning context. But because the College Board provides syllabi, sample tasks, evaluation rubrics, and instructional support materials for AP courses, many aspects of the critical student/task matchup are in place before the examination.

Use Case 4: Large-scale educational surveys

In large-scale educational surveys such as the National Assessment for Educational Progress (NAEP), samples of students are administered assessments to survey achievement across jurisdictions and to support research on its correlates. It is similar to Use Case 2 in that tasks are administered to students about whom relatively little is known. But it differs as to the intended claims: Not inferences about individuals, but about distributions of performance, relationships with demographic and educational background variables, and patterns of performance on some more complex tasks. In the last of these, rich work products such as log files of students' actions are obtained, providing evidence about the processes by which students perform tasks: their choices, the way they use tools, the steps they take, where they run into problems, and so on (for examples, see the 2014 NAEP Technology and Engineering Literacy (TEL) assessment: http://www.nationsreportcard.gov/tel_2014/).

Chapter 3
Evidence-Centered Assessment Design

Abstract *Design patterns* are tools to support task authoring under an evidence-centered approach to assessment design (ECD). This chapter reviews the basic concepts of ECD, focusing on evidentiary arguments. It defines the attributes of design patterns, and shows the roles they play in creating tasks around valid assessment arguments.

The *design patterns* described here support designing tasks to assess students' capabilities to carry out model-based reasoning as sketched above. They build on tools and concepts from an evidence-centered approach to assessment design (ECD; Mislevy, Steinberg, & Almond, 2003; Mislevy & Riconscente, 2006). Messick (1994) lays out the essential narrative of assessment design, saying that we

> ...begin by asking what complex of knowledge, skills, or other attributes should be assessed, presumably because they are tied to explicit or implicit objectives of instruction or are otherwise valued by society. Next, what behaviors or performances should reveal those constructs, and what tasks or situations should elicit those behaviors? (p. 16).

ECD distinguishes layers at which activities and structures appear in assessment, to create operational processes that instantiate an assessment argument (described later in this section). Table 3.1 summarizes the layers. *Design patterns* are tools for working in the Domain Modeling layer, where research and experience about the domains and skills of interest that have been marshaled in Domain Analysis will be organized in the form of assessment arguments.

To show how *design patterns* support this work, we extend Toulmin's (1958) general argument structure to assessment arguments. By conceptualizing assessment as a form of argument, we can use *design patterns* to support design choices in terms of its elements. Further discussion on how assessment arguments are then instantiated in the machinery of operational assessments—stimulus materials, scoring procedures, measurement models, delivery systems, and so on—appear in Almond, Steinberg, and Mislevy (2002), Mislevy, Steinberg, and Almond (2003) and Riconscente, Mislevy, and Corrigan (2015).

© The Author(s) 2017
R.J. Mislevy et al., *Assessing Model-Based Reasoning using Evidence-Centered Design*, SpringerBriefs in Statistics, DOI 10.1007/978-3-319-52246-3_3

3.1 Assessment Arguments

An evidentiary argument is constructed through a series of logically connected propositions that are supported by data via warrants, and are subject to alternative explanations (Toulmin, 1958). Figure 3.1 applies Toulmin's argument structure to educational assessment. The claims concern aspects of students' proficiency—what they know or can do in various settings. Data consist of (1) their observed behaviors in particular task situations, (2) the salient features of those tasks, and (3) other relevant information the assessment user may have about the relationship between the student and the task situation, such as personal or instructional experience. Warrants posit how responses in situations with the noted features depend on the proficiency (or proficiencies) we intend to assess. Some conception of knowledge and its acquisition—i.e., a psychological perspective—is the source of warrants, and shapes the nature of claims an assessment is meant to support, and the tasks and data needed to evidence them (Mislevy, 2003, 2006).

In our case, research on model-based reasoning provides the warrants. The research cited in the previous section suggests how students with certain kinds of knowledge and capabilities for reasoning through particular models would be apt to do in what kinds of task situations. Alternative explanations for poor performance are deficits in the knowledge or skills that are required to carry out a task but are not focal to the claim, such as familiarity with the computer interface used in a simulation-based investigation. These are "construct irrelevant" requirements (Messick 1989).

This assessment-argument structure applies directly to stand-alone tasks such as multiple-choice items and short answer tasks. The data concerning the task situation

Table 3.1 Layers of evidence-centered design for assessments

Layer	Role	Key entities
Domain analysis	Gather substantive information about the domain of interest that has direct implications for assessment; how knowledge is constructed, acquired, used, and communicated	Domain concepts, terminology, tools, knowledge representations, analyses, situations of use, patterns of interaction
Domain modeling	Express assessment argument in narrative form based on information from domain analysis	Knowledge, skills and abilities; characteristic and variable task features, potential work products, potential observations
Conceptual assessment framework	Express assessment argument in structures and specifications for tasks and tests, evaluation procedures, measurement models	Student, evidence, and task models; student, observable, and task variables; rubrics; measurement models; test assembly specifications; PADI templates and task specifications
Assessment implementation	Implement assessment, including presentation-ready tasks and calibrated measurement models	Task materials (including all materials, tools, affordances); pilot test data to hone evaluation procedures and fit measurement models
Assessment delivery	Coordinate interactions of students and tasks: task-and-test-level scoring; reporting	Tasks as presented; work products as created; scores as evaluated

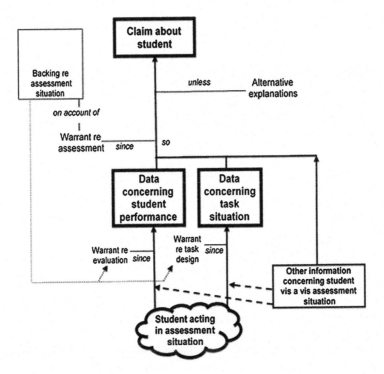

Fig. 3.1 An extended Toulmin diagram for assessment arguments

are "built in at the factory" by the task author. The student responds to that situation; that is, says, does, or makes something—i.e., work products (which may include the processes and intermediate stages of work). Work products are evaluated to produce the data concerning the student's performance, through a rubric, a scoring guide, or an automated scoring procedure.

The structure applies to more complex and interactive tasks as well. Here the task is not a fixed situation, but an evolving sequence of situations, which may be modified in response to the student's actions (and sometimes in accordance with the situation's own logic, such as a patient's disease changing over time in a medical diagnosis task). A chain of structures like Fig. 3.1 will result, as suggested in Fig. 3.2. Work products that are produced can now include the actions at each moment, in the situation as it is at that moment. The features of performance are thus evaluated, and indeed, only make sense, in light of the situation the student has worked herself into, and sometimes in light of her previous actions.

Evaluating a performance can be more complicated in these situations. A traditional way to evaluate interactive performances is for a human to evaluate them, looking for evidence in actions as they occur and the situation unfolds. Certain features of performance may be sought, counted, noted for presence or absence, or judged holistically. In computer-based simulations, automated scoring procedures can be used to evaluate students' work in these less-structured tasks.

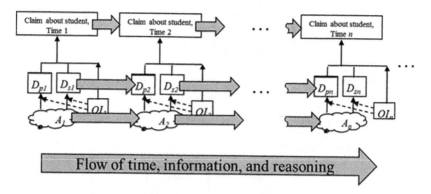

Fig. 3.2 Assessment argument structure for an evolving task

Details are beyond this presentation, but Bejar, Mislevy, Rupp, and Zhang (2016) provide an overview of automated scoring procedures for complex assessment tasks. Examples in STEM appear in DiCerbo et al. (2015), Gobert, Sao Pedro, Baker, Toto, and Montalvo (2012) and Rupp et al. (2012). Discussions of techniques from data mining to further explore potential observations in complex, interactive, simulation spaces appear in DiCerbo et al. (2015) and Gibson and Clarke-Midura (2015).

The point is that the underlying logic for task design and scoring is essentially the same. In the *design patterns* described in the following sections, features of task situations can be designed-in, or they can be recognized as they emerge in ongoing interactions. Similarly, potential observations can be predetermined features of predetermined work products, or they can be features of an ongoing performance, evaluated in light of the situations that exist at the time they are made. The Genetics and Robotics Boxes provide some illustrations.

3.2   Design Patterns

While Toulmin diagrams provide support for understanding the structure of an assessment argument, *design patterns* provide support for creating its substance. Table 3.2 defines the key attributes of a PADI *design pattern* and specifies which component of the assessment argument it concerns. Their meanings and uses become clearer in the sections discussing the *design patterns*. The most detailed discussion is the first, Model Formation, in Chap. 5. It is foundational to understanding model-based reasoning as well as design patterns.

Design patterns are intentionally broad and non-technical: "centered around some knowledge, skills, or abilities (KSAs), a *design pattern* is meant to offer a variety of approaches that can be used to get evidence about that knowledge or skill, organized in such a way as to lead toward the more technical work of designing particular tasks" (Mislevy & Riconscente, 2006, p. 72). Because *design patterns* do

Table 3.2 Basic design pattern attributes, definitions, and corresponding assessment argument components

Attribute	Definition	Assessment argument component
Name	Short name for the design pattern	
Summary	Brief description of the family of tasks implied by the design pattern	
Rationale	Nature of the KSA of interest and how it is manifest. Concisely articulates the theoretical connection between the data to be collected and the claims to be made	Warrant
Focal knowledge, skills, and abilities (KSAs)	The primary knowledge/skill/abilities targeted by this design pattern	Claim
Additional KSAs	Other knowledge/skills/abilities that may be required by tasks motivated by this design pattern	Claim if relevant; alternative explanation if irrelevant
Potential work products	Things students say, do, or make that can provide evidence about the focal knowledge/skills/abilities	Data concerning students' actions
Potential observations	Features of work products that encapsulate evidence about focal KSAs	Data concerning students' actions
Characteristic features	Aspects of assessment situations which are likely to evoke the desired evidence	Data concerning situation
Variable features	Aspects of assessment situations that can be varied in order to control difficulty or target emphasis on various aspects of KSAs	Data concerning situation
Examples	Samples of tasks that instantiate this design pattern	
References	Research, applications, or experience relevant to task design under this design pattern	Backing

not include the technical specifics of domain content, psychometrics, or task delivery (these considerations come into play in the next layer, the Conceptual Assessment Framework, or CAF), they provide a common planning space for various experts that may be involved in the assessment design process, such as curriculum developers, item writers, psychometricians, teachers, and domain specialists.

Using *design patterns* to create assessment tasks provides benefits in validity, generativity, and reusability. First, validity is strengthened as tasks inherit the backing and rationale of the *design patterns* from which they were generated. Creating a *design pattern* for some aspect of proficiency requires articulating the components of the assessment argument, including the line of reasoning that explicates why certain kinds of data can offer evidence about that proficiency. The *design pattern* is connected to backing, or the research and experience that ground, the argument. Laying out the argument frame before developing specific tasks in their particulars helps ground the interpretation of test scores. *Design patterns* remain a resource for subsequent task development, serving as explicit and sharable backing for new tasks in the same application or other applications that address the same areas.

A second benefit is generativity. Because *design patterns* organize experience across past research and projects that all address the assessment of some targeted aspects of learning, they support the creation of new tasks grounded in an established and explicit line of reasoning. Organizing *design patterns* around aspects of learning, especially ones that are difficult to assess, helps a task designer get started much more quickly: Scaffolding is provided about the shape of the argument, approaches that have been used in the past, and examples of tasks that illustrate the ideas.

A third benefit of *design patterns* is reusability. A *design pattern* encapsulates key results of work from the Domain Analysis stage and reflects the form of an assessment argument. As such it helps to structure a designer's work in both Domain Analysis and Domain Modeling. The same *design pattern* can motivate a great many tasks in different areas and at different levels of proficiency, all revolving around the same hard-to-measure aspects of, say, scientific inquiry; their particulars can be detailed with the content, purposes, constraints, and resources of the assessment at hand.

It may be noted in passing that the basic assessment design pattern structure can be augmented in various ways to incorporate additional information or considerations into the design process. Two examples are the following:

1. *Accessibility and test accommodations.* Construct-irrelevant knowledge, skills, or capabilities to access, interact with, and respond task situations become alternative explanations for poor performance, as they can become the source of a student's difficulties. Design patterns can be augmented to help provide each student with forms of tasks for which construct-irrelevant demands are minimized (Haertel, Haydel DeBarger, Villalba, Hamel, & Mitman Colker, 2010; Hansen, Mislevy, Steinberg, Lee, & Forer, 2005; Mislevy et al., 2013; Rose, Murray, & Gravel, 2012).

2. *Learning progressions.* Learning progressions are "hypothesized descriptions of the successively more sophisticated ways student thinking about how an important domain of knowledge or practice develops as children learn about and investigate that domain over an appropriate span of time" (Corcoran, Mosher, & Rogat, 2009, p. 37). Design patterns can be augmented to support the creation of tasks that provide information about students' proficiencies with respect to a learning progression by adding structure and relations across three key attributes (West et al., 2012; Zalles, Haertel, & Mislevy, 2010). Elaborating design patterns in these terms is relevant to assessing model-based reasoning for two reasons. First, the NGSS are organized in terms of grade bands that describe typical patterns of increasing proficiency, which are amenable to the progression-like ordering described above. Second, model-based reasoning is among the aspects of science learning that researchers have investigated from the perspective of learning progressions (e.g., Schwartz et al., 2009). Creating PADI *design patterns* for the Schwartz et al. (2009) learning progressions would be straightforward, and provide additional support designers creating tasks that target learning progressions or at grade bands.

Chapter 4
Design Patterns for Model-Based Reasoning

Abstract The aspects of model-based reasoning serve as the Focal knowledge, skills and abilities (KSAs) of the *design patterns*. They highlight distinct aspects of model-based reasoning in a way that supports either focused tasks (building on one or a few design patterns) or more extensive investigations (building jointly on several *design patterns*). This chapter overviews the design-pattern perspective on assessing model-based reasoning, as a prelude to the next chapters that look more closely at each aspect. A table charts the correspondence between the aspects addressed in the design patterns and practices in the Next Generation Science Standards.

Distinguishable aspects of model-based reasoning, involving different, though overlapping, kinds of knowledge and processes, must be coordinated in investigations. The *design patterns* presented here highlight distinct aspects of model-based reasoning but in a way that supports designing either focused tasks (building on one or a few *design patterns*) or more integrated investigations (building jointly on several design patterns).

The six *design patterns* described in Chaps. 5–10 address particular aspects of model-based reasoning. Their essential interaction in practice is the concern of the Model-Based Inquiry *design pattern* in Chap. 11. The Appendix summarizes them in tabular form.[1] To emphasize their utility beyond isolated aspects of model-based reasoning but also their interplay in practice, the examples will show how the design patterns apply in both focused and integrative tasks.

The aspects of model-based reasoning listed in Table 4.1 serve as the Focal KSAs of the *design patterns* presented here. They are meant to guide task design across the range of scientific models which can differ in content and detail. Content and level of detail are therefore Variable Features of tasks in all of these *design patterns*, and familiarity with the content and representational forms associated with particular models is a corresponding Additional KSA of each *design pattern*. What will be common to all tasks motivated by a given *design pattern*, however, will be the Characteristic Features—those features that are essential in a problem setting if

---

[1]Online read-only versions are available at http://design-drk.padi.sri.com/padi/do/NodeAction?state=listNodes&NODE_TYPE=PARADIGM_TYPE.

R.J. Mislevy et al., *Assessing Model-Based Reasoning using Evidence-Centered Design*, SpringerBriefs in Statistics, DOI 10.1007/978-3-319-52246-3_4

Table 4.1 Mapping of NGSS performance expectations for NGSS Science and Engineering Practice 2: Developing and using models

Grade	NGSS Science Practice	Design Pattern (DP)	Focal KSA(s)
K-2	Distinguish between a model and the actual object, process, and/or events the model represents	Model evaluation	Ability to differentiate between a model and a real-world phenomenon, including identifying how they align and how they differ, and identifying limitations of the model
K-2	Compare models to identify common features and differences	Model evaluation	Ability to compare two or more models, including identifying how they differ in their mapping to the real-world phenomenon, identifying the trade-offs of the model, and comparing the model's appropriateness for a specified purpose
K-2	Develop and/or use a model to represent amounts, relationships, relative scales (bigger, smaller), and/or patterns in the natural and designed world (s)	Model formation	Aligns to the DP as a whole
		Model use	Ability to reason through the concepts and relationships of a given model to make explanations, predictions and conjectures
K-2	Develop a simple model based on evidence to represent a proposed object or tool	Model formation	Aligns to the DP as a whole
3-5	Identify limitations of models	Model evaluation	Ability to differentiate between a model and a real-world phenomenon, including identifying how they align and how they differ, and identifying limitations of the model
3-5	Collaboratively develop and/or revise a model based on evidence that shows the relationship among variables for frequent and regularly occurring events	Model formation	Aligns to the DP as a whole
		Model revision	Aligns to the DP as a whole
3-5	Develop a model using an analogy, example, or abstract representation to describe a scientific principle or design solution	Model formation	Aligns to the DP as a whole
3-5	Develop and/or use models to describe and/or predict phenomena	Model formation	Aligns to the DP as a whole
		Model use	Aligns to the DP as a whole
3-5	Develop a diagram or simple physical prototype to convey a proposed object, tool or process	Model articulation	Ability to transform information between qualitative and/or quantitative systems associated with scientific phenomena
3-5	Use a model to test cause-and-effect relationships or interactions concerning the functioning of a natural or designed system	Model use	Ability to reason through the concepts and relationships of a given model to test explanations, predictions and/or conjectures
6-8	Evaluate limitations of a model for a proposed object or tool	Model evaluation	Ability to differentiate between a model and a real-world phenomenon, including identifying how they align and how they differ, and identifying limitations of the model

(continued)

Table 4.1 (continued)

Grade	NGSS Science Practice	Design Pattern (DP)	Focal KSA(s)
6–8	Develop or modify a model, based on evidence, to match what happens if a variable or component of a system is changed	Model formation	Aligns to the DP as a whole
6–8		Model revision	Aligns to the DP as a whole
6–8	Use and/or develop a model of simple systems with uncertain and less predictable factors	Model formation	Aligns to the DP as a whole
6–8		Model use	Aligns to the DP as a whole
6–8	Develop and/or revise a model to show the relationship among variables, including those that are not observable but predict observable phenomena	Model formation	Aligns to the DP as a whole
6–8		Model revision	Aligns to the DP as a whole
6–8	Develop a model to describe unobservable mechanisms	Model formation	Aligns to the DP as a whole
6–8	Develop and/or use a model to generate data to test ideas about phenomena in natural or designed systems, including those representing inputs and outputs, and those at unobservable scales	Model formation	Aligns to the DP as a whole
6–8		Model use	Aligns to the DP as a whole
9–12	Evaluate merits and limitations of two different models of the same proposed tool, process, mechanism or system in order to select or revise a model that best fits the evidence or design criteria	Model evaluation	Ability to compare two or more models, including identifying how they differ in their mapping to the real-world phenomena, identifying the trade-offs of the model, and comparing the models' appropriateness for a specified purpose
9–12	Design a test of a model to ascertain its reliability	Model evaluation	Ability to design a method for testing a model
9–12	Develop, revise, and/or use a model based on evidence to illustrate and/or predict the relationship between systems or between components of a system	Model formation	Aligns to the DP as a whole
9–12		Model use	Aligns to the DP as a whole
9–12		Model revision	Aligns to the DP as a whole
9–12	Develop and/or use multiple types of models to provide mechanistic accounts and/or predict phenomena, and move flexibly between model types based on merits and limitations	Model formation	Aligns to the DP as a whole
9–12		Model use	Aligns to the DP as a whole
9–12	Develop a complex model that allows for manipulation and testing of a proposed process or system	Model formation	Aligns to the DP as a whole
9–12	Develop and/or use a model (including mathematical or computational) to generate data to support explanations, predict phenomena, analyze systems, and/or solve problems	Model formation	Aligns to the DP as a whole
9–12		Model use	Aligns to the DP as a whole

it is to evoke evidence about the Focal KSA. To assess Model Revision, for example, there must be an existing model, information that conflicts with it, and a need to revise the model to accommodate the discordant information. On the other hand, such tasks may vary as to the scientific model of interest and other features; examples include the following:

- the existing model was provided, or generated by the student;
- the task is focused solely on model revision, or model revision is a multiply occurring aspect to be evaluated in the context of a larger investigation;
- students are working independently or in groups; and
- the students' work takes place in hands-on investigations, open-ended written responses, oral presentations, or multiple-choice tasks.

These possibilities are highlighted for the designer in the Variable Task Features and Potential Work Products attributes.

Implications of a key presumption of the *design patterns* can now be seen for the contents of their attributes and how they are used in practice. The *design patterns* address aspects of reasoning, but model-based reasoning is always about something. These are *general* patterns to support creating *specific* tasks: tasks that involve reasoning with particular models in particular circumstances. The terms, concepts, representational forms, and procedures associated with a model will always be intimately involved with tasks created from these *design patterns*. Thus substantive knowledge of the model(s) at issue will be an Additional KSA in every *design pattern* that follows. This alerts the task designer to important design choices concerning the interplay among the model-based reasoning that is targeted by a task, knowledge of the elements and processes of the particular models, and knowledge of the substantive aspects of whatever situation is presented.

Suppose, for example, the desired focus is metacognitive: assessing students' capabilities to manage and carry out the interacting phases of a model-based investigation. A design choice is using very simple models that are accessible to a wide range of students, or models that are more substantial but known to be familiar to the students who will be assessed. In these cases, substantive knowledge as an Additional KSA is *by design* unlikely to be a source of construct irrelevant variance; that is, students are unlikely to perform poorly simply because they aren't familiar with the underlying model.

Alternatively, if the desired focus is assessing students' capabilities at carrying out reasoning steps with a particular model that is also the focus of attention, as when that model is a current target of instruction, then the demands for knowledge of that model can be higher. The assessor wants to know if the student can carry out reasoning with that model, so failure due to lack of familiarity with that model is now construct relevant.

Another Variable Task Feature involves several Additional KSAs, and holds implications for decisions about work products and observable variables: whether the task is to be carried out by a group a students or by students working independently. When tasks are carried out by a group, the Characteristic Features, Focal

KSAs, Work Products, and so on concerning the targeted aspects of model-based reasoning are still pertinent. However, group tasks induce Additional KSAs concerning skills of communication, interaction, explanation, and persuasion that can also be targets of inference. Work products, potential observations, scaffolding, and other design decisions concerning collaboration would be addressed in the same integrated activity as model-based reasoning. A task developer could draw jointly upon multiple *design patterns* such as the Model Formation and "Participating in Collaborative Scientific Inquiry." Collaboration *design patterns* are not included in this suite, but an early example appears in Mislevy et al. (2003). Developing a suite of design patterns for collaborative skills that reflects recent research (e.g., Hesse, Care, Buder, Sassenberg, & Griffin, 2015) would be a worthy project for a team of cognitive scientists and educators.

The design patterns (DPs) listed here were developed to highlight the main aspects of model-based reasoning. Concepts from different parts of model-based reasoning can be mapped on to the design patterns and the focal KSAs listed in the design pattern. For example, Table 4.1 shows a mapping of the NGSS Science and Engineering Practice 2: Developing and Using Models concepts (NGSS). For each standard, there is a corresponding *design pattern*. Some standards map to a *design pattern* as a whole. For example, the grade K-2 standard of "Develop a simple model based on evidence to represent a proposed object or tool" is aligned to the Model Formation DP. The Focal KSAs in the Model Formation DP further break down what it means to develop a simple model.

In other cases the, the standard aligns to a particular FKSA. For example, the grade 3–5 standard of "Identify limitations of models" can be mapped to the FKSAs of "Ability to differentiate between a model and a real world phenomena, including identifying how they align and how they differ, and identifying limitations of the model" within the Model Evaluation DP. The other features of the DP can be used to help identify salient features of a task that would allow for the measurement of this standard/FKSA.

Chapter 5
Model Formation

Abstract Model Formation begins by selecting and assembling model elements to establish a correspondence with some situation, often in the real world or a corpus of data. The Model Formation *design pattern* addresses features of this contextualized process that are similar across contexts and models. Design choices concerning the knowledge and skill that will be encompassed, variable features of tasks, potential work products and observations.

A scientific model is an abstract system of entities, relationships, and processes. Every particular use of a model begins by selecting and assembling model elements to establish a correspondence with particular circumstances—often real-world situations, but also possibly the entities, processes, and relationships in other models. We call this aspect of model-based reasoning Model Formation. (The NGSS calls it Developing Models; other terms are Model Building, Model Construction, and Model Instantiation.) This section presents a *design pattern* for assessing model formation, whether in focused tasks or as an integrated aspect of model-based reasoning. It is summarized in the first column of the Appendix.

5.1 Rationale, Focal KSAs, and Characteristic Task Features

Even though model formation is inherently about instantiating particular models in particular contexts, the Model Formation *design pattern,* like the others, doesn't specify a particular model or context. We are not proposing, however, that model formation is a decontextualized ability, independent of particular models and contexts.[1] Rather, the *design pattern* addresses features of the contextualized activities that are similar across contexts and models. These features are similar

---

[1]It is the case, however, that an individual can develop through experience a generalized schema for how and when to use models, and procedures and strategies for using them, which can be called upon to guide reasoning with new models and in new contexts (Perkins and Salomon 1989).

© The Author(s) 2017
R.J. Mislevy et al., *Assessing Model-Based Reasoning using Evidence-Centered Design*, SpringerBriefs in Statistics, DOI 10.1007/978-3-319-52246-3_5

31

enough that making them explicit and organizing them around assessment arguments supports designing tasks that evidence this aspect of model-based reasoning across different contexts and models.

A task supported by the Model Formation *design pattern* involves a real-world situation, such as a system, a problem setting or corpus of data, and a purpose for formulating a model. These are Characteristic Features of model formation tasks.

The Focal KSAs (FKSAs) of this *design pattern* are aspects of the model formation process, as instantiated in the context of given situations and models:

- Ability to relate elements of the model to elements of the situation, and vice versa.
- Ability to describe the situation through the entities and relationships of the model.
- Ability to pose relevant questions about the situation to inform the construction of the model.
- Ability to identify which aspects of the situation to address and which to omit, including the scope and grain-size of model.
- Decision-making regarding scope and grain-size of a model, as appropriate to the intended use of the model.

Depending on the purpose of the assessment, a designer may focus on some of these aspects more than others, and address them separately or as an ensemble. The Variable Task Features section discusses how choosing features of tasks can elicit one or another aspect of the model formation process. For now, we note that the first two FKAs listed above highlight the correspondence between elements of the model and features of the situation. The last three highlight the correspondence among the model, the situation, and the purpose of modeling. This entails identifying which aspects of the situation are relevant and which can be safely ignored, and justifying the accuracy needed for the purpose. The Potential Work Products attribute shows that these aspects of reasoning may be elicited explicitly, inferred from intermediate work or think-aloud solutions, or lie implicit in the student's formulated model.

Boxes Genetics-2 and Robotics-2 illustrate some of the attributes from the Model Formation design pattern in the context of the running examples introduced previously.

Box Genetics-2. Model-Based Reasoning Tasks in Genetics: Model Formation

Assessments of a students' proficiency in using models appear throughout Stewart and Hafner's genetics course. The same *design pattern* can be used for different assessments by modifying the model in question as well as other Variable Features. As the course progresses, the assessments may be more focused on other elements of model-based reasoning, but elements from model use are still involved. The following is a task that can be used at the beginning of the course when the focus is mainly on model use:

Complete the Punnett square to show the possible outcomes of a cross of a heterozygous father with a widow's peak with a homozygous mother with a widow's peak.

		Father's genotype?	
		Possible sperm?	Possible sperm?
Mother's genotype?	Possible egg?		
	Possible egg?		

A. What fraction of offspring would have a widow's peak?
B. What fraction of offspring would not have a widow's peak?
(http://www.cccoe.net/genetics/punnett4.html)

The model in this example is a co-dominance model for how alleles combine. Students are asked to reason through this model to apply it to make predictions regarding the offspring. Moreover, they must do so using the Punnett square representation, choosing the correct parent traits to cross and performing the crosses correctly. They then must be able to interpret the results in terms of possible traits of the offspring.

Notice that the answers that students give to problems A and B are dependent on them filling in the Punnett Square representational form appropriately. One possible Observable Variable is the joint correctness of the square and the question responses. A more nuanced rubric could first evaluate the correctness of the square and then evaluate students' question responses conditional on the way they completed the square. Even if they did not fill in the square correctly, they can still demonstrate some appropriate reasoning through the model by providing answers that are consistent with their square. For example, mistaking the relationship for simple dominance would lead to incorrect predictions, but reasoning from the Punnett square under this presumption does indicate appropriate steps of model use.

Providing the Punnett square is a design choice that supported students in using an appropriate tool for some steps in reasoning through the model. Not providing it would then provide evidence about whether a student could recall and use this representational form to reason through the inheritance model. Separate Observable Variables would be called for, as recalling the form is not equivalent to being able to reason through the model.

Box Robotics-2. Model Formation
Students start the hill-climbing challenge in the simulation space. After they have built (and revised as necessary) the electrical circuit for the motor in the simulation space, they must now construct a gearbox, place it and the attached motor on a chassis, and connect it to the drive wheels of the simulated rover. The task begins with a default 1:1 ratio from the motor to the wheels (The assessment designers know the rover will not have enough torque to climb the ramp; this design choice ensures that the student will need

to revise the model). Forming the model requires manipulation of the Gazebo tools, to position and attach the components of the model so they are connected and the motor will turn the wheels when the students turns it on. Whether the simulated rover can climb the simulated hill remains to be determined in subsequent model use and model evaluation.

There is ongoing, implicit, assessment in the student's formation of the simulation-model. If the components are not correctly connected, the simulation provides feedback in a visible form: the wheels do not turn, and further manipulation is required. As noted further below, additional assessment could be layered onto this phase of activity.

The Focal KSA in this instance of Model Formation is building a simulation model–specifically, of a device with certain components, connections, and characteristics, to produce the motive behavior. Additional KSAs in this phase of the task are being able to use the Gazebo tools and representations to construct and run assemblies of components. Substantive knowledge about circuits, gearboxes, and motors to the extent that they are needed in this phase of the task are also Additional KSAs. By the time the student has reached this task in the MOOC, however, the instruction material and Gazebo exercises will be familiar and usually not a source of difficulty. The primary difficulty is model formation, in the form of constructing the simulation model; hence model formation as particularized to this rover-model space, in this context, for this purpose. Figure R3 illustrates this process using the model editor in the simulation software.

The Characteristic Feature of the model formation aspect of the simulation phase is the requirement to assemble a model of an artifact to be produced. This is an engineering-centered application of model-based reasoning.

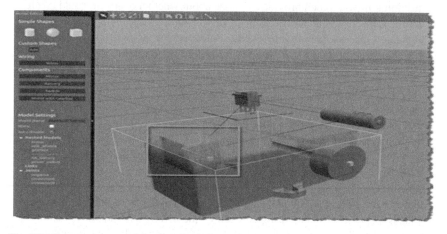

Fig. R3 Constructing a model using simulation software

Note that the same artifact, scientific models, and challenge appear in the simulation phase and the physical-construction phase of the task. *A key Variable Task Feature differs: Whether the construction takes place in the simulation world or the physical world.* There are obvious similarities, intentionally so, so that students appreciate the rapid and inexpensive testing cycles they can carry out with the simulation. But different Additional KSAs are required to act in the simulation world (familiarity with the interfaces and affordances) and physical world (ability to manipulate tools and sometimes-stubborn components). By intention and by design, the simulation and physical modeling will help students come to realize an important fact of modelling in engineering: the behavior of a model in the model space need not be identical to the behavior of the corresponding artifact in the real world.

The work product in the simulation phase is the ongoing model as the student constructs it—that is, at any given point in work, there is a computer log file is being accumulated that describes the constructed model. Gazebo can compute how it will behave (or fail to behave) when the students carries out acts such as moving it and switching it on. This implicit evaluation provides feedback to the students as to whether she has formed a functional simulated rover. These are fast, implicit, informal assessments, which don't even look like assessments as we are used to thinking about them. They can be understood nevertheless in terms of the assessment argument structure and its application to model formation aspect of model-based reasoning, as an instance of particularizing the attributes of the design pattern.

One could build more explicit assessment for the simulation activities in model formation. Using the same log file, one could automatically report a number of other observations: number of attempts, locations where components were successfully or successfully connected, presence of common errors, and so on. Additional work products could be collected, for example by asking a student to write an explanation of how he constructed the model. Potential Observations that can be obtained from an explanation include its completeness, indications of misconceptions, and its accuracy overall and with respect to targeted features.

After a number of cycles of model formation, evaluation, and revision in the simulation space (discussed in upcoming sections), the student assembles physical components to build a rover in the real world. The goal is create an instantiation of the successful simulation model. The focal KSA is again model formation, now in a different yet strongly related problem. Understanding of components and the relationships, and their assembly are again Additional KSAs, which again have been supported through the MOOC. New Additional KSAs are understanding of the physical components and being able to manipulate them—measuring and matching gear placements on an axle, for example, and tightening the gear hub with a hex wrench. The corresponding actions in the simulation were pointing and clicking. Here they are controlled actions in physical space. Although the

Additional KSA of understanding of these system components is required in both cases, different Additional KSAs are required to form the model in the two phases, namely proficiency with the simulation tools and objects and proficiency with physical tools and objects. A student can struggle forming the model in one phase but not the other because of different levels of familiarity in the two environments.

Now there is no log-file Work Product being accumulated automatically. The simplest work product to obtain is the final rover the student assembles. Potential observations associated with this work product are the degree and particular ways the assembled rover corresponds to the simulation model, the articulation of the components, and their functioning (whether it can actually climb the hill comes later, in model evaluation). More extensive potential work products would be traces of the student's actions, as observed by, say, a teacher, or a more transportable and persistent video capture. Potential observations again can include number of attempts, presence and kinds of errors, and mismatches between the simulated rover and the physical rover. Additional work products can be collected, such as interviews of students' explanations of how they assembled the components and problems they may have had, or their notes in a lab book.

5.2 Additional KSAs

Additional KSAs are other aspects of knowledge that may or may not be involved in a model formation task at the discretion of the task designer, in accordance with the context and intended use of the task. They call attention to design choices that will either intentionally impose or minimize demands on particular models and on other knowledge, skills, and abilities. Primary among Additional KSAs is knowledge of the model(s) that will be involved. In some applications, the designer may want to assess students' ability to form models of a given type when it is known that the students are familiar with the elements of the model. In others, assessing both knowledge of the elements of a model and being able to instantiate it in a given setting will be of interest jointly.

For example, Marshall (1993, 1995) asked students to select an arithmetic schema from the five they have been studying (Change, Group, Compare, Vary, and Restate), then map elements of a word problem to its slots (Fig. 5.1). A teacher using Marshall's curriculum is implicitly conditioning her inferences on the knowledge that these models and these representational forms are familiar to the students. This focuses the evidentiary value of the task on model formation using these models. Using the same tasks in a large-scale survey assessment confounds knowledge of the schemas and representations with the ability to match them to real-world situations.

INSTRUCTIONS: Choose the one diagram below that fits this story problem. Move the arrow into the diagram you have selected and click the mouse button.

Dan Robinson recently drove 215 miles from San Diego to Santa Barbara to see his parents. When he arrived at his parents', he noticed that the odometer of his car registered 45631 miles. What was the odometer reading before he made the trip?

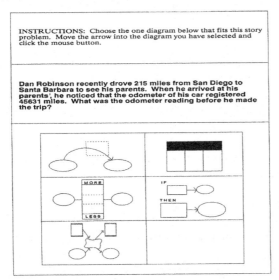

ASSESSING SCHEMA KNOWLEDGE

INSTRUCTIONS: Choose the one diagram below that fits this story problem. Move the arrow into the diagram you have selected and click the mouse button.

Dan Robinson recently drove 215 miles from San Diego to Santa Barbara to see his parents. When he arrived at his parents', he noticed that the odometer of his car registered 45631 miles. What was the odometer reading before he made the trip?

FIG. 7.6.   Constraints task from SPS: diagrams.

INSTRUCTIONS: Identify the parts of the problem that belong in the diagram. Move the arrow over each part. Click and release the mouse button. Drag the dotted rectangle into the diagram, and click the mouse button again when you have positioned the rectangle correctly in the diagram. If you make a mistake, return to the problem and repeat the process. When you are finished, move the arrow into the OKAY box and click the mouse button.

Harry the computer programmer accidentally erased some of his computer programs while he was hurrying to finish work one Friday afternoon. Much to his dismay, when he returned to work on Monday, he discovered that only 24 programs of his original 92 programs had survived. How many computer programs had been destroyed?

Fig. 5.1 Two tasks for selecting and filling in an appropriate schema (Marshall 1993, p. 167)

As mentioned, knowledge of the model and/or content domain is always involved in model formation. Domain-specific knowledge structures, principles, procedures, and heuristics are at the heart of experts' model formation. Newell and Simon (1972) called these "strong methods" for problem solving, in contrast to domain-independent "weak methods" such as means-ends analysis and trial-and-error. Seeing a physics problem in terms of Newton's Third Law or a genetics problem in terms of a genes-code-for-proteins schema are examples of domain-specific heuristics and domain-specific explanatory schemas (Duncan,

2006). Being able to bring such strategies to bear is important in any assessment where learning in the domain, with its models and principles, is at issue.

A critical design choice with respect to domain-specific knowledge in a model-formation task thus depends on two related questions–*neither of which is visible in the task itself*. Is the domain knowledge a part of the construct to be assessed? What will the user of the assessment know about the examinee's experience with the domain knowledge? These are some familiar cases that illustrate the design choices:

- Suppose the focus is the process of model-formation in conjunction with a particular model. Both knowing the components and structures of the models *and* being able to articulate them with real-world situations is the instructional goal—hence, the construct to be assessed. This situation often occurs in formative assessment and summative assessment that is linked to instruction.
- Suppose the focus is the process of model formation per se, to be assessed using a model known to be familiar to examinees. The challenge of the task is now on identifying, matching, and relating aspects of the situation to the elements and relationships of a familiar model. This known familiarity rules out difficulties due to declarative and representational aspects of the model. Different students could be assessed with different models in otherwise comparable tasks, in order to ensure that all of the students are familiar with the model in their tasks.
- Suppose the focus is again on the process of model formation per se, but the task is to be administered to students about whom the designer or user knows very little. The designer may employ a substantive model that is expected to be familiar with almost all examinees—say Marshall's arithmetic schema tasks, for middle-school students who are familiar with the arithmetic but may or may not be able to carry out the construction and mapping processes that connect the models to real-world situations. Note that the evidentiary value of a task depends on what knowledge is required in the task *and* the user's knowledge of the examinees' backgrounds with those elements.

Familiarity with the task type and stimulus materials is another Additional KSA in tasks addressing model formation and the other model-based reasoning aspects that follow. For a student who is not familiar with a task type, irrelevant sources of difficulty can include what a problem is asking, how the information is presented, how responses are to be made, and what is expected in a response.

It is important for students to learn to solve near-transfer problems (Bransford & Schwartz, 1999) at an early stage of learning. However, unfamiliar tasks that yield to familiar models with novel mappings—far-transfer problems—are important in the long run (Clement, 2000). It is by extending their experience across a situations with diverse surface features that students begin to organize their knowledge around underlying principles.

Interfaces, tools, representational forms, and symbol systems that appear in tasks can be essential to success, whether they appear as stimuli, are required in solution processes, or are needed to produce work products. A task designer interested in model formation with a given model will want to use only tools and representations students are familiar with, in order to avoid construct irrelevant sources of difficulty from this source.

Although it is not a focus of this discussion, we note that other enabling knowledge and skills such as language, vision, and mobility that may be required in a task are also Additional KSAs. These demands may need to be minimized or circumvented to improve the accessibility of tasks for students with special needs (Hansen, Mislevy, Steinberg, Lee, & Forer, 2005; also see Mislevy et al., 2013, and Rose, Murray, & Gravel, 2012, on how the principles of Universal Design for Learning (UDL) can be formally incorporated into the ECD framework and in *design patterns*).

5.3 Variable Task Features

There is an important connection between Task Features, over which a task designer has considerable control, and Focal and Additional KSAs, which are the aspects of examinees' capabilities a task is meant to elicit. Choices about Variable Task Features address task features that increase or decrease the demand for both Focal and Additional KSAs. These choices should be made purposefully. The following relationships between Variable Task Features and KSAs help a task developer recognize and think through design issues.

A Variable Feature alluded to in the discussion of Additional KSAs is the familiarity of the context and the problem format. Using familiar task for students first encountering a model reduces cognitive load, and allows the students to focus on working with the model rather than on figuring out the task. But unfamiliar tasks and contexts are needed to develop, then assess, students' capabilities to form models in far transfer situations (Redish, 2004).

Whether a task is "near transfer" or "far transfer" cannot be determined just by looking at a task. It depends on knowing the relationship between the task and the instructional and experiential history of a student. Data in the form of other information about the relationship between the student and the task (Fig. 2.1) thus play an important role in determining how the designer should manipulate this Variable Feature. If the user does not know this relationship (it often is not known in a "drop-in-from-the-sky" test), the evidentiary value of a student's response is degraded by alternative explanations. Is a performance misleadingly good only because this problem type was already familiar to her? Does another student fare poorly with a task that is usually easy because he had never encountered that problem type before?

Two Variable Features that affect task difficulty can be manipulated to create easy tasks, challenging tasks for advanced students, or tasks that lie somewhere between. They are the *complexity of the model* and the *complexity of the situation* to be modeled. Other things being equal, the need to use a more complex model makes a problem harder. Complexity features in a *model* include the number of variables or elements, the complexity of their interrelations, the number of representations required, and whether multiple models need to be used and integrated (see the Model Elaboration *design pattern*). Complexity features in a *situation* include the number and variety of elements in the real-world situation, the presence of extraneous information, and the degree to which elements have been stylized in order to make their identification and subsequent model formation easier. Difficulty can be increased by having more possible alternatives as to what to include in a model, or how detailed or extensive a model must be to meet the task's goal.

Tasks can also vary as to the extent to which students are familiar with the context, in order to avoid extraneous knowledge requirements (as discussed with Additional KSAs) or to intentionally incorporate requirements for substantive knowledge. Incorporating demands for substantive knowledge can be desirable either because the user already knows that examinees are familiar with it or because that knowledge is itself a target of inference along with being able to use it to form models.

Tasks also can vary with regard to the amount of scaffolding provided. Songer, Kelsey, and Gotwals (2009) created a design pattern for writing tasks with decreasing levels of scaffolding to go along with the BioKIDS instructional program, to support students' moving up a learning progression for scientific explanation. Marshall's schema selection tasks are scaffolded, as befits students who are learning to work with these models. Figure 5.2 is a task with less scaffolding. Heller and Heller (2001) developed problems that encourage students to (a) consider physics concepts in the context of real objects in the real world; (b) view problem-solving as a series of decisions; and (c) use the fundamental concepts of physics to qualitatively analyze a problem before manipulating formulas.

The Hellers also reduced inappropriate scaffolding by avoiding "trigger words" in their problem statements, such as "starting from rest" and "inclined plane." These terms activate physics schemas—usually the correct ones in textbook exercises. But educators want students to develop associations grounded in underlying principles

---

You've been hired as a technical consultant to the Minneapolis police department to design a radar detector-proof device that measures the speed of vehicles. (i.e. one that does not rely on sending out a radar signal that the car can detect.) You decide to employ the fact that a moving car emits a variety of characteristic sounds. Your idea is to make a very small and low device to be placed in the center of the road that will pick out a specific frequency emitted by the car as it approaches and then measure the change in that frequency as the car moves off in the other direction. The device will then send the initial and final frequencies to its microprocessor, and then use this data to compute the speed of the vehicle. You are currently in the process of writing a program for the chip in your new device. To complete the program, you need a formula that determines the speed of the car using the data received by the microprocessor. You may also include in your formula any physical constants that you might need. Because your reputation as a designer is on the line, you realize that you'll need to find ways to check the validity of your formula, even though it contains no numbers.

Fig. 5.2 Example of a "context rich" problem (Heller and Heller 2001, p. 104)

Population Dynamics: Predator/Prey Relationship

As a member of the International Committee for the Protection of Threatened and Endangered Animals (ICPTEA), you have been asked to respond to a subcommittee's report that there has been a rapid decline in the snowshoe hare population over the past four years. The major predator of the snowshoe hare is the lynx. In order to prevent the continued decline of the hare population, the subcommittee has proposed reducing the lynx population.

Previous research has shown that the snowshoe hare survives by eating the sparse plant material growing in the cold climate of Canada, and that the hare is capable of rapid population growth due to its high birthrate. The lynx has a much lower birthrate than the hare.

You have found the following data on the population levels of each species in a given region over a 28 year period (MacLulick, 1937). The population of hares is given in thousands, and the population of lynx is given in hundreds.

Time elapsed years	Population of snowshoe hare (thousands)	Population of lynx (hundreds)
0	20	10
2	55	15
4	65	55
6	95	60
8	55	20
10	5	15
12	15	10
14	50	60
16	75	60
18	20	10
20	25	5
22	50	25
24	70	40
26	30	25
28	15	5

To develop a clearer understanding of the research data in the table, plot the data on a line graph. Make sure that the axes are clearly labeled. Designate the snowshoe hare populations with a dot (.) and the lynx populations with an (x).

    1. Using the data in the table and your graph, explain the relationship, if any, between the populations of lynx and snowshoe hares.

    2. Write a response to the members of the subcommittee stating whether you support or reject the proposal to reduce the lynx population. Explain your decision using information you have obtained from the table of data and your graph.

Fig. 5.3 Predator/prey task

rather than surface features of problem statements, and to form models in situations beyond stylized teaching examples.

    The example shown in Fig. 5.3 further illustrates choices concerning Variable Features in a model formation task, here concerning the Predator/Prey relationship.[2] Students are given a table with data for the hare and lynx populations in an area, and are told the lynx is a predator of the hare. They are asked to determine the

---

[2]Downloaded July 31, 2007 from //pals.sri.com/tasks/5-8/ME406/directs.html. The Council of Chief State School Officers (CCSSO) contributed this task to the Performance Assessment Links in Science (PALS) library.

relationship (i.e., formulate a model) between the population sizes. Tables and graphs are required, introducing the Additional KSA of familiarity with these representational forms. There is scaffolding for data analysis using a coordinate graph—which ameliorates weaknesses with graphing techniques as an alternative explanation for poor performance—but no scaffolding for model formation. This combination of choices about what knowledge to support focuses the evidentiary value of the task on model formation rather than analytic methodology.

5.4   Potential Work Products and Potential Observations

Because the cognitive processes of model formation are not directly visible, an assessment must use the things students say, do, or make as evidence. The forms that contain the evidence are the work products of a task. Model Formation tasks can be designed to elicit a variety of Work Products, each with its own resource requirements, knowledge demands, aspects of thinking it can provide evidence about, and quality of information obtained.

A related design choice is determining which aspects of work products to discern and evaluate from a performance as captured in a work product. These are the Observable Variables. *Design patterns* provide support to a task developer by suggesting kinds of qualities that can be the basis for defining Observable Variables, or Potential Observations. Some Observable Variables attend to a given work product and others address relationships among work products. Potential Observations in a *design pattern* may be supplemented with rubrics, which, broadly construed, are the processes—algorithms, instructions, or guidelines—that people or machines apply to Work Products to determine the values of Observable Variables.

The Potential Work Products attribute of this *design pattern* suggests things students could say, do, or make that hold evidence about their model formation capabilities. A model formation task could produce Work Products associated with the final model they generate, the process taken to produce it (e.g., a log file of actions), or explanations and justifications of the model. A final model takes the form of knowledge representations, such as coordinated diagrams, a physical construction, or a system of equations with explanations of variables and relationships in terms of the target situation. A Work Product in a model formation task could be the selection of a model from among a given set, such as Marshall's schema selection tasks; a constructed model in a constrained and therefore scaffolded work space, such as Marshall's fill-in-a-schema tasks; a freely-generated model in some representational form; or a physical model that embodies the key elements and relationships.

With the availability of computer-based task administration, a wide variety of response forms can be used for students to express a model in constructive and open-ended ways that lend themselves to automated scoring (Bejar, Mislevy, Rupp, & Zhang, 2016; Scalise & Gifford, 2006). When the form of the Work Product is produced with a technology-based tool, Additional KSAs are introduced with

Fig. 5.4 Stock and flow
diagram for a model of
population growth

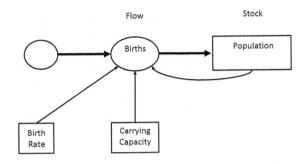

respect to the familiarity with the representational form and whatever interfaces are
required. The use of such tools can however be intimately related to understanding
certain kinds of models, such as software programs used in interactive data analysis
and modeling interactive systems. The STELLA package (Richmond, 2005) provides tools for building dynamic models, working back and forth among diagrams,
equations, data, and graphs of interactive systems. Figure 5.4 is a stock and flow
diagram similar to the ones the STELLA program uses for a simple model of
population growth (Allen, Kling, & van der Pluijm, 2005), corresponding to the
equation

$$\text{Births} = \text{Birth Rate} * \text{Population} * (1 - (\text{Population}/\text{Carrying Capacity})).$$

Work Products that concern a *final product* provide direct evidence about the
quality of the outcome of the model formation process, and hold clues as to which
elements of the process succeeded and which did not. A student's STELLA model
for the population growth problem, for example, conveys evidence about whether
she has considered feedback in the system.

Work Products that concern the formation *process* can include questions the
students pose to themselves or to others, notes taken during construction, and traces
of the steps taken during in formulating the model (e.g., notes, diagrams, computer
logs of actions). These Work Products may be written, spoken, or interactions with
an interface. They may be captured on computer logs, video tape, transcribed, or
only heard by the assessor. They may be responses to explicit directives (e.g.,
answers to multiple-choice questions), answers to informal questions posed during
instruction, or unstructured comments as from think-aloud solutions. Compared to
final solution Work Products, process-oriented Work Products can provide more
direct evidence for metacognitive aspects of model formation, and add support for
formative feedback.

Several kinds of Observable Variables can be evaluated from these various Work
Products. Regarding final models, Potential Observations include the quality and
accuracy of the final model, incorporating aspects such as the degree to which
targeted aspects of the situation are represented in the model, the efficiency of the
models and representations, whether extraneous elements are included in the model,
and the appropriateness of the precision used for the goal of the task. As an

example, Kindfield's (1999) study of diagrams that examinees used to explain crossover in meiosis supports defining an Observable Variable based on the inclusion of extraneous elements in a model. Novices' drawings were often more complete and better proportioned than experts', but the experts' diagrams tended to show only the salient features, and the relationships important to the problem were rendered just with the accuracy needed to solve the problem. That is, the experts' diagrams were more efficacious than those of the novices.

When the Work Product is a functioning or runnable model, whether virtually as with STELLA or physically as in the robotics example, potential features to evaluate address its behavior in the circumstances it is meant to approximate. Does it capture the key elements? Does it function properly within certain ranges but fail outside others? Since many engineering approximations have this property, are the failures outside the scope of the problem?

Potential Observations regarding process can address time efficiency, the quality of self-monitoring reflections, and the properties of intermediate models. For example, were there many restarts or scrapped work, as opposed to incremental improvements? How much time was spent in planning before a first provisional solution was produced? For tasks involving mental models, rapid correct solutions as opposed to slow correct solutions provide evidence for automatized model formation, a characteristic of expert–like knowledge (Kalyuga, 2006).

When the Work Product is students' explanations of their models, Potential Observations include an awareness of considerations involved in choosing model elements, the quality and accuracy of the model-situation relationships and the degree of accuracy of modeling (overall and/or with respect to specified features). In assessments where domain knowledge is construct-relevant, Observable Variables can be based on whether or not, and if so how effectively, students employed domain-specific heuristics and explanatory schemas.

In the genetics running example, model formation is targeted when the students are given two sets of genes and must determine the probability for offspring to possess each type of trait (Sect. 6.2 and Box Genetics-2). A model for the relationships is required. If Punnett Squares have been previously introduced, the students must first recognize this as a situation in which this representation can be used, and then use it to explain how their inheritance models, expressed in this form, fit the data. The students' model formation can be evaluated with regard to the accuracy of their instantiation of the Punnett Square. (The quality and accuracy of their performance segues into Model Use, as Model Formation and Model Use are integrated in this task.)

Alternatively, suppose the students have not been exposed to Punnett squares and the instructor wants the students to formulate an information-equivalent model expressed in terms of the relationships among alleles and phenotypes. This instructor can present the students with information on different sets of parents' genes and traits, and the genes and traits of their offspring. Using this information, the students must formulate how genes are combined and come up with the idea of the simple dominance model. One Work Product would be the representation of the final model, which can be evaluated for correctness and completeness.

Rubrics describe the process by which features of Work Products are discerned and evaluated as Observable Variables, which in turn convey information on some aspects of the student's process or product in formulating a model. In the PALS predator/prey task, students produce a graph of the relationship between the hare and the lynx populations and a written explanation of the relationship. These are the two Work Products. The rubric shown as Fig. 5.5 assigns a score on a 1 to 4 scale, where a 1 signifies a wholly inadequate graph in terms of a list of targeted properties, and a 4 signifies a correct depiction of the relationship in a syntactically correct graph.

In a more complex example, Azevedo and Cromley (2004) assessed the quality of the models students constructed to explain a diagram of the human circulatory system after studying in different hypermedia environments. Their rubric shown in Fig. 5.6 summarizes a student's model in terms of increasingly more accurate and sophisticated understandings of the components and processes of the circulatory system. The Observable Variable is 'level of explanation,' as evaluated from a Work Product in the form of a transcript from a talk-aloud explanation. This rubric derives from research on progressive understandings of the circulatory system (e.g.,

Rubric	
NS	No attempt to graph (labels, numbers or plotting of any of the data onto the grid) is present. No adequate analysis (demonstration of understanding of the relationship between the snowshoe hare and lynx population by mentioning any of the aspects of the graph) is given.
1	Student demonstrates limited understanding of graphing and limited understanding of the relationship between the snowshoe hare and lynx populations. Example: An attempt to graph (labels, numbers or plotting of any of the data onto the grid) may be present. No adequate analysis (demonstration of understanding of the relationship between the snowshoe hare and lynx population by mentioning any aspects of the graph) is given. Student may indicate something about the data or, the trends or, the labels of their graph.
2	Student demonstrates some understanding of graphing and some knowledge of the relationship between the snowshoe hare and lynx populations. For example, the graph is constructed, the trends are accurate, and some data for the hare or the lynx is correctly plotted (but may be in thousands, not hundreds) or missing. The answer suggests that the student does not understand the relationship between the snowshoe hare and lynx populations OR the graph is not constructed correctly and plotted accurately, but the answer does demonstrate that the student understands the relationship between the snowshoe hare and lynx populations by mentioning at least one aspect of the graph.
3	Student demonstrates adequate understanding of graphing and adequate knowledge of the relationship between the snowshoe hare and lynx populations. Example: The graph is constructed correctly and data for the hare is plotted adequately (no more than three data points misplotted). Data for the lynx may be plotted in thousands, not hundreds, but has been adequately plotted (no more than three data points misplotted). An answer that demonstrates the understanding of the relationship between the snowshoe hare and lynx populations by mentioning at least two aspects of the graph is present (i.e., and increase in hare population leads to a lynx population increase; there is a delay in the change of the populations of snowshoe hare and lynx; there are ten times as many hares as lynx).
4	Student demonstrates a high level of understanding of graphing and a high level of knowledge of the relationship between the snowshoe hare and lynx populations. The graph is constructed correctly and data for the hare and lynx is plotted accurately. The difference in scale of the hare data and the lynx data is accurate. The correct analysis of the data is made by noting the three aspects of the graph. The delay in the change of the populations is noted (i.e., when the lynx population increases, years later the snowshoe hare population begins to decrease, and when the snowshoe hare population decreases, years later the lynx population begins to decrease).

Fig. 5.5 Rubric for Item 1 of the PALS predator/prey task

Circulatory System Model – Rubric

1. No understanding
2. Basic Global Concepts
   - blood circulates
3. Global Concepts with Purpose
   - blood circulates
   - describes "purpose" - oxygen/nutrient transport
4. Single Loop – Basic
   - blood circulates
   - heart as pump
   - vessels (arteries/veins) transport
5. Single Loop with Purpose
   - blood circulates
   - heart as pump
   - vessels (arteries/veins) transport
   - describe "purpose" - oxygen/nutrient transport
6. Single Loop - Advanced
   - blood circulates
   - heart as pump
   - vessels (arteries/veins) transport
   - describe "purpose" – oxygen/nutrient transport
   - mentions one of the following: electrical system, transport functions of blood, details of blood cells
7. Single Loop with Lungs
   - blood circulates
   - heart as pump
   - vessels (arteries/veins) transport
   - mentions lungs as a "stop" along the way
   - describe "purpose" – oxygen/nutrient transport
8. Single Loop with Lungs - Advanced
   - blood circulates
   - heart as pump
   - vessels (arteries/veins) transport
   - mentions Lungs as a "stop" along the way
   - describe "purpose" – oxygen/nutrient transport
   - mentions one of the following: electrical system, transport functions of blood, details of blood cells

9. Double Loop Concept
   - blood circulates
   - heart as pump
   - vessels (arteries/veins) transport
   - describes "purpose" - oxygen/nutrient transport
   - mentions separate pulmonary and systemic systems
   - mentions importance of lungs
10. Double Loop – Basic
   - blood circulates
   - heart as pump
   - vessels (arteries/veins) transport
   - describe "purpose" - oxygen/nutrient transport
   - describes loop: heart - body - heart - lungs - heart
11. Double Loop – Detailed
   - blood circulates
   - heart as pump
   - vessels (arteries/veins) transport
   - describe "purpose" - oxygen/nutrient transport
   - describes loop: heart - body - heart - lungs –heart
   - structural details described: names vessels, describes flow through valves
12. Double Loop - Advanced
   - blood circulates
   - heart as pump
   - vessels (arteries/veins) transport
   - describe "purpose" - oxygen/nutrient transport
   - describes loop: heart - body - heart - lungs - heart
   - structural details described: names vessels, describes flow through valves
   - mentions one of the following: electrical system, transport functions of blood, details of blood cell

Fig. 5.6 Necessary features for evaluating models of the circulatory system

Chi, 2005). Note that the same backing could be used to create alternatives for multiple-choice items or to develop rubrics for working models of the circulatory system.

5.5 Considerations for Larger Investigations

The Model Formation *design pattern* is meant to support the authoring of both tasks that focus solely on model formation and tasks that include model formation as a part of a larger activity. A task could entail model formation then model use, for example, or formation-use-evaluation, or full investigation that engages all phases of inquiry.

A full investigation could be scaffolded to distinguish model formation phases for the student, or the student could need to recognize when and how to form models. In the latter case, recognizing and managing phases calls upon the knowledge addressed in the final Model-Based Inquiry *design pattern*. Alternatively, a task that walks a student through the stages minimizes the need for managing the inquiry phases, and thus does not provide evidence about this capability, but does evoke evidence specific to phases. By determining task features in such ways, the designer can tune the evidentiary value of a task to targeted aspects of model-based reasoning.

To make sense of extended performances in a larger task context, it can be useful to notice and evaluate model formation, as well as other aspects of model-based reasoning, as they take place within the evolving context. For example, a student may formulate an inappropriate model, but reason through that model correctly. The points made above regarding Characteristic Features, Potential Work Products, and Potential Observations still hold.

The task developer has a degree of control over how explicitly to elicit evidence about aspects of reasoning by means of design choices about Work Products. In designing a simulation-based assessment in dental hygiene, Mislevy, Steinberg, Breyer, Almond, and Johnson (2002) found that the trace of actions—despite being a rich and detailed Work Product—did not convey students' intermediate mental products such as identification of cues, generation of hypotheses, and selection of tests to explore conjectures. They added a Work Product in the form of an insurance coding sheet, similar to those now integral to actual practice, on which the examinee would indicate hypotheses based on cues from various sources of information and justify information-gathering actions with hypotheses or as standards-of-care.

5.6   Some Connections to Other Design Patterns

The Model Formation *design pattern* can be viewed as a subpart of the Model-Based Inquiry. To design and create extended performances in an inquiry investigation, the developer can use the Model-Based Inquiry *design pattern* to coordinate the overall activity and the constituent *design patterns* (such as Model Formation) to guide inquiry phases, Work Products, and Evaluation Procedures for the multiple phases of activity.

Many familiar tasks combine Model Formation with Model Use, the *design pattern* addressed next. A problem context is given, and a solution is required: The student must formulate a model and reason through it to obtain a solution. A task developer can choose among (1) evaluating the product of both aspects of model-based reasoning, so that evidence is evoked about either combined success or failure somewhere along the way, (2) obtaining discernable (though dependent) evidence about formation and use by structuring Work Products that distinguish the stages, or (3) obtaining a rich Work Product such as a talk-aloud solution, traces of solution steps in a log file, or intermediate products and then seeking evidence by applying rubrics that address both model formation and model use.

The Model Formation *design pattern* overlaps with Model Elaboration and Model Revision. As an aspect of model-based reasoning, model elaboration focuses on combining or making additions to a model, such as embedding it in a larger system or adding elements or submodels, or connecting to another model to form multiple, multilevel, or composite models. Model revision is a kind of model formation, but with a focus on responding to shortcomings from a given model as prompted by feedback from the environment such as incorrect predictions or lack of fit to data.

It is possible to create finer-grained *design patterns* for model formation, such as having *design patterns* for mental models and for deliberative modeling. The *design pattern* presented here is meant to be broadly useful across domain areas, educational levels, and types of assessments. It thus offers less specific support for any particular area, level, or assessment type. More specialized *design patterns* could be developed in any of these respects, to provide stronger support for designers who need to develop tasks for these situations.

Chapter 6
Model Use

Abstract Model use is reasoning through the entities, relationships, and processes of a given model to provide explanations, make predictions, or fill in gaps with respect to real-world situations or summary data about real-world situations. The Model Use *design pattern* describes characteristic and variable features of tasks for eliciting this thinking, and work products and observations to capture and interpret the evidence that results.

Model use is reasoning through the entities, relationships, and processes of a model to provide explanations, make predictions, or fill in gaps with respect to data or real-world situations. Model use is a necessary component in building, testing, and revising models. Some instruction and some assessment of model use with given models focus mainly on reasoning through the relationships within a model, while other instruction and assessment require model use in coordination with model formation, testing, and revising.

6.1 Rationale, Focal KSAs, and Characteristic Task Features

Figure 1.1 shows model use as making inferences about the real-world situation originally depicted at the lower left, through the relationships of the model in the middle plane—that is, in terms of the version of the situation reconceived through the model, at the lower right. Such thinking might be "run" in one's head as a mental model or supported by tools or external representations (e.g., a mechanical model or a computer simulation) as suggested by the representational forms and operations at the top of the figure.

Assessment tasks can highlight various kinds of this reasoning. Among the most important is *explanation* of a physical situation, which entails articulating the relationships among observations and events in terms of the underlying concepts,

R.J. Mislevy et al., *Assessing Model-Based Reasoning using Evidence-Centered Design*, SpringerBriefs in Statistics, DOI 10.1007/978-3-319-52246-3_6

principles, and relationships of the model. For example, in order to give a "complete" explanation in a task from the Earth-Moon-Sun curriculum,

> ...students have to put the relevant elements together into phenomenon-object-motion (POM) charts, which include an explanation using both text and diagrams, and articulate the relationship between their celestial motion model and the phenomenon in question (often using props such as inflatable globes, Styrofoam balls, and light sources) (Stewart, Passmore, Cartier, Rudolphn. & Donovan, 2005, p. 161).

Making predictions, constructing retrodictions (i.e., what might have happened previously for things to be as they are now?), and filling in missing information about a real-world situation are also varieties of model use; one must reason through the relationships of the model to infer entities or circumstances in the future or the past, or not immediately observable. The DP differentiates *making* prediction and explanations from *testing* them. While both require reasoning through the model, testing additionally requires reasoning about alternative explanations.

One can distinguish qualitative reasoning through model concepts from using symbol systems and knowledge representations. Larkin (1983) showed that experts solved physics problems first by building an understanding of the situation in terms of the underlying principles and relationships, and only then proceeding to develop the systems of equations to solve the problems. Hestenes (1987) argued that the emphasis placed on mathematical methods in college physics instruction and assessment slights conceptual understanding and biases students toward a formula-based approach rather than a model-based approach. Figure 6.1. is a typical task of the kind Hestenes deems insufficient. The formula-based approach does not produce the desired deeper understanding of the underlying physics.

In response, Hestenes, Wells, and Swackhamer (1992) developed the Force Concept Inventory (FCI), tasks that present situations that require only qualitative reasoning through fundamental concepts. Figure 6.2. is an example. Two similar assessments focusing on qualitative reasoning through central models are the Force and Motion Conceptual Evaluation (FCME; Thornton and Sokoloff, 1998), which addresses kinematics at a more advanced level, and the Test About Particles in a Gas (TAP; Novick & Nussbaum, 1981), which concerns the particulate nature and

---

A projectile is fired horizontally from a flare gun located 45.0 m above the ground. The projectile's speed as it leaves the gun is 250 m/s.

a) How long does the projectile remain in the air?
b) What horizontal distance does the projectile travel before striking the ground?
c) What is its speed as it strikes the ground?
d) If the projectile were simply dropped from a height of 45.0 m, instead of fired horizontally from that height, how much time would it take to reach the ground? How does this compare with your answer to part (a)?

Fig. 6.1 A formula-based model use task

USE THE STATEMENT AND FIGURE BELOW TO ANSWER THE NEXT FOUR
QUESTIONS (8 THROUGH 11).
The figure depicts a hockey puck sliding with constant speed v_o in a straight line from point "a"
to point "b" on a frictionless horizontal surface. Forces exerted by the air are negligible. You are
looking down on the puck. When the puck reaches point "b," it receives a swift horizontal kick
in the direction of the heavy print arrow. Had the puck been at rest at point "b," then the kick
would have set the puck in horizontal motion with a speed of v_k in the direction of the kick.

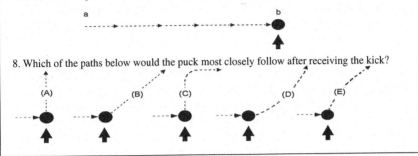

8. Which of the paths below would the puck most closely follow after receiving the kick?

Fig. 6.2 A task from the Force Concept Inventory (from Hestenes, et al., 1992)

behavior of gases. Quantitative reasoning generally follows qualitative reasoning in
practice, and cycling between the two is common. As a Variable Feature, model use
tasks can be designed to focus on just one or the other, or their interaction.

The Focal KSAs at the heart of the Model Use *design pattern*, then, are the
capabilities to make explanations, predictions, retrodictions, and fill in missing
elements in the context of some model(s) and situation(s). This encompasses
qualitative or quantitative manipulations, or both, as required. Tasks based on this
design pattern share Characteristic Features: a situation and one or more models
that the students apply to reason about it. The model(s) may be only provisional,
because model evaluation and model revision may need to follow. Box Robotics-3
discusses aspects of Model Use and potential assessment of model use in the
running robotics example.

Box Robotics-3. Model Use

Model Use in the robotics task concerns reasoning through a constructed
model—in the first phase, simulation models, in the latter phase, physical
models—to anticipate the rover's behavior. When the student completes the
circuit, the motor should run at its base speed, the power should be trans-
ported through the gearbox, and the wheels should turn at a speed determined
by the gearing ratios. This is *cognitive* model use. Model Use is closely
connected with Model Formation as the MOOC provides support so that a
correctly formed simulation model or physical model is likely to move. There
is thus a great deal of scaffolding for the cognitive model use, which is a
choice for a key variable task feature.

The goal is for the rover to move, and in fact to climb the ramp, but this will depend on the gear ratios, weight distribution, and type of wheels the student chooses. Success, and, more importantly, failures that lead to revisions that produce success, will come in cycles of formation, use, evaluation, and revision.

Whereas reasoning through a mental model is purely cognitive model use; "running the model" means reasoning through the model elements, relations, and processes to an expected outcome. With engineering models, one can also run the model operationally; in this case, completing the circuit to start the motor and produce behavior. In the simulation space, the behavior is simulated as well, calculated through the computer model that was produced in Model Formation. In addition to observing the simulated movement on the screen, the student can also request live production of a number of graphs relating aspects of the behavior to each other and to time (Fig. R4). These will become important in Model Evaluation. Note that being able to use the graphing tool and understand the graphical representations are Additional KSAs. A student who understood the underlying scientific models and how they operate in a rover would still struggle in the simulation phase if he could not access or understand the use of these affordances.

In the physical space, the behavior is produced by the assembled components. There are no automated graphs, but a learning goal is that students' experience with the plots in the simulation phase will add insight to their interpretations of behaviors they see in terms of those relationships.

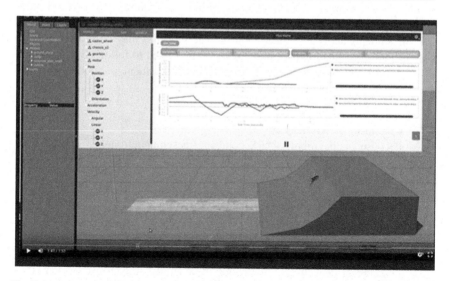

Fig. R4 The graphing tool plots multiple hill-climbing simulations for comparison. The *green* and *red* trial outperforms the previous trial (*purple* and *blue*). Comparing wheel rotation to forward motion can help diagnose wheel slip problems

Model Use is thus largely implicit in the robotics task. After the first attempt, there is evidence in a revised model, in the student's choice of gear ratios, weight distribution, and/or wheel types would be predicted through a correct model to produce behavior closer to the targeted behavior. For example, if on attempt t the student had seen the wheels were turning too fast and didn't have enough torque to climb the ramp as it became steeper, the gear ratio the student chooses for attempt $t + 1$ should have a lower ratio, not a higher ratio. This evidence about model use is a by-product of model revision, as discussed in Chap. 10.

Suppose the task designer wanted more explicit evidence about students' cognitive model use. Design choices to this end could require predictions or explanations of a rover's behavior before it is run; for example what will be its speed compared to the base motor speed? What will be its speed compared to the previous attempt? Will it make it farther up the hill? In each case, why do you think so? What equations are using to ground your prediction?

6.2   Additional KSAs

As with Model Formation, Additional KSAs that may be involved in a task assessing model use include familiarity with the concepts, entities, relationships in a given model, and associated tools and representational forms. That is, both the declarative knowledge that is necessary to support reasoning through the model and whatever supports are required for apprehending, interacting with, and responding to a task must also be taken into account when drawing inferences from students' performances. Demands for such ancillary skills can enhance a task's evidentiary value, as when knowledge of representation software is known to be familiar to the examinee and can be used to support their reasoning. Or such demands can degrade a task's evidentiary value, as when examinees perform poorly due to a lack of necessary but ancillary capabilities (Wiley & Haertel, 1996).

A user may be interested in all of these KSAs jointly, or focus the evidentiary value of a task selectively in light of what else is known about the relationship between the examinees and the task requirements. For example, task might call for prediction from a model known to be familiar, or solving a familiar kind of problem with a new model. Box Genetics-2, a continuation of the Genetics Toolkit Example, is an instance of the latter. Students fill in a now-familiar Punnett square for a co-dominance model just after the model has been introduced. As with other *design patterns* for model-based reasoning, model-using may be assessed in a task focusing on this aspect alone—model and data given, appropriateness presumed, at least provisionally—or as part of a larger task.

Tasks for assessing model use often require model formation. A design choice for a task developer is whether to assess them jointly, separately, or sequentially. The following section discusses tradeoffs among these alternatives.

Box Genetics-2. Model-Based Reasoning Tasks in Genetics: Model Formation

Assessments of a students' proficiency in using models appear throughout Stewart and Hafner's genetics course. The same *design pattern* can be used for different assessments by modifying the model in question as well as other Variable Features. As the course progresses, the assessments may be more focused on other elements of model-based reasoning, but elements from model use are still involved. The following is a task that can be used at the beginning of the course when the focus is mainly on model use:

Complete the Punnett square to show the possible outcomes of a cross of a heterozygous father with a widow's peak with a homozygous mother with a widow's peak.

		Father's genotype?	
		Possible sperm?	Possible sperm?
Mother's genotype?	Possible egg?		
	Possible egg?		

A. What fraction of offspring would have a widow's peak?
B. What fraction of offspring would not have a widow's peak?
(http://www.cccoe.net/genetics/punnett4.html)

The model in this example is a co-dominance model for how alleles combine. Students are asked to reason through this model to apply it to make predictions regarding the offspring. Moreover, they must do so using the Punnett square representation, choosing the correct parent traits to cross and performing the crosses correctly. They then must be able to interpret the results in terms of possible traits of the offspring.

Notice that the answers that students give to problems A and B are dependent on them filling in the Punnett Square representational form appropriately. One possible Observable Variable is the joint correctness of the square and the question responses. A more nuanced rubric could first evaluate the correctness of the square and then evaluate students' question responses conditional on the way they completed the square. Even if they did not fill in the square correctly, they can still demonstrate some appropriate reasoning through the model by providing answers that are consistent with their square. For example, mistaking the relationship for simple dominance would lead to incorrect predictions, but reasoning from the Punnett square under this presumption does indicate appropriate steps of model use.

Providing the Punnett square is a design choice that supported students in using an appropriate tool for some steps in reasoning through the model. Not providing it would then provide evidence about whether a student could recall and use this representational form to reason through the inheritance model. Separate Observable Variables would be called for, as recalling the form is not equivalent to being able to reason through the model.

6.3 Variable Task Features

Variable Task Features include, as in the other *design patterns*, the model(s) at issue, students' familiarity with the model and task type, the complexity of the model(s), the amount and kinds of scaffolding, whether work is completed in a group or independently, and whether the targeted model use is embedded in a larger activity.

Additional variable task features are whether the data or the model are provided or are generated by the student in a previous phase of a task. A tradeoff arises: If model and data are provided, the developer can focus the evidentiary value of the task on whether the student can carry out the reasoning through that model. However, little information would then be obtained about whether the student can manage the inquiry activities that characterize real-world model use. This decision can be appropriate when specified aspects of model use are the focus. Alternatively, suppose model use is assessed in a less structured manner, in which the student must collect data, formulate a model, then reason through the model. Now difficulties in earlier stages of work may prevent the student from providing evidence about using the models of interest. On the other hand, more evidence is obtained about managing the phases of inquiry. A compromise design option is to stage an investigation in phases such that when students have trouble forming an appropriate model, they are provided hints or scaffolding so that they can then carry out model use with the intended model.

6.4 Potential Work Products and Potential Observations

Student Work Products that can be captured in model use include explanations, predictions, retrodictions, and filled-in information in the form of verbal, written, diagrammatic, symbolic, or physical media (see Scalise & Gifford, 2006, on computer-based formats for Work Products that are amenable to automated scoring). Let the term "solution" encompass hypotheses, predictions, explanations, and/or missing elements of a real-world situation. Three basic kinds of Work Product can be obtained to provide evidence about aspects of model use: The solution itself, traces of the solution, and explanations of the solution.

The solution itself. In traditional large-scale assessments, this can take the form of selecting a solution from offered alternatives, as with multiple-choice items. Alternatively, the student may construct the solution through representational forms: simply a word or number, diagram or chart, or a more elaborated description of preconditions, possible causes of an event or predictions about possible outcomes. The forms of solution may be generated by the student, or, as scaffolding, the student may complete given representational forms, possibly partially filled in.

The FCI example in Fig. 6.2. shows that a thoughtfully constructed multiple-choice task can provide a great deal of information about students' thinking. The distractors are designed to elicit common misconceptions. The curved options here look like the parabolic paths of horizontally propelled objects that are subject to gravitational force—paths that are, in fact, correct answers to other FCI tasks that depict physically different situations. These distractors appeal to students whose understanding of forces is still at a surface level.

Traces of the solution. Traces of model using can be tracked to capture intermediate steps, key strokes and action-selections in computer-based solutions, and think-aloud protocols. Martin and VanLehn's (1995) OLAE system for solving kinematics problems, for example, records each step of a student's solution, including restarts. These Work Products hold increasing value as tasks become more complex.

Explanations of the solution. A student can be asked to provide a written or oral description of a solution, how it was obtained, and its rationale. A presentation to other students is a formal and structured example. In contrast to solution traces, an explanation requires verbalizing steps, strategies, and rationales of the model use. Possible qualities to discern and evaluate, or Potential Observations, are the completeness and the accuracy of the reasoning of a prediction or explanation.

When the Work Product takes the form of a final solution, correctness and accuracy are usually of interest. In simple problems, this may suffice. In more complex problems, however, much thinking and many steps—hence, much potential evidence—takes place that may not be apparent in the solution alone. It can then be of interest to examine the steps taken in reasoning through the model and to evaluate the process in such terms as appropriateness, efficiency, systematicity, quality of strategy, and effectiveness of procedures. Evaluating Observable Variables such as the trace of a solution requires a method for detecting and summarizing its salient qualities. From an explanation Work Product, the Observable Variables concern the student's capability to express these qualities. Requiring explanations additionally benefits instruction by making the steps of model use overt, thus amenable to student reflection and supportive of metacognitive skills.

Choices about Observable Variables are linked to choices about Work Products; some Work Products support a given Potential Observation and others do not. If evidence is desired about capabilities to build a diagrammatic representation from a verbal description, then a labeled diagram is an appropriate Work Product to elicit, and Observable Variables pertain to its adequacy and correctness. If evidence is desired about associating elements of equations to aspects of diagrams, students can be asked to generate or select equations, and rubrics are needed to evaluate adequacy and correctness. If evidence about misconceptions is desired, a variety of Work Products could be elicited, as long as both the design of the task and the evaluation procedures made it possible to evoke a misconception and capture evidence about it: Rubrics can be used to identify misconceptions from open-ended Work Products, while closed-ended Work Products such as multiple-choice responses on the FCI present options that reflect particular misconceptions.

When the Work Products in a given task can include explanations, it is also possible to evaluate the quality of students' reasoning about their own reasoning. This kind of Work Product and Observable Variable draws attention to metacognition—specifically, the quality with which students are monitoring and evaluating their own use of the model. Asking students to evaluate their own or other students' reasoning as they use models brings this thinking into their awareness, and supports learning as well as assessment.

6.5   Some Connections with Other Design Patterns

As noted above, tasks that combine model formation and model use are common: A student is presented a real-world situation and asked to solve a problem, provide an explanation, or make a prediction or retrodiction. In simple problems, these aspects of model-based reasoning are difficult to individuate. In FCI multiple-choice tasks, for example, the only Work Product is the response choice. The distractors have been constructed so a correct response suggests the student both formulated and reasoned through the correct model, while an incorrect choice suggests the student has formulated and reasoned through a model based on a particular misconception.

One can, however, capture evidence separately for model formation and model use by requiring Work Products that specifically express the model being formed, then the reasoning through that model. Observable Variables based on the multiple Work Products—each motivated by the corresponding *design pattern*—can then be developed.

Most tasks that address model evaluation and model revision also involve model use. In both cases, it is necessary to reason through a provisional model in order to compare its predictions with a situation. For model evaluation, the model in focus is the model being evaluated. In model revision, the models being reasoned through are alternatives to a given model, to see if their predictions accord with the situation better than the present model.

Chapter 7
Model Elaboration

Abstract Model elaboration focuses on combining or extending a model, such as embedding it in a larger system, adding elements or submodels, or connecting it with other models to form multilevel or composite models. The Model Elaboration *design pattern* describes characteristic and variable features of tasks, and potential work products and observations.

In scientific practice, new information sometime requires major revisions to our current understanding of a situation. Other times, information moves us to extend the instantiation of a current model being used in a project or investigation, or to integrate multiple portions of familiar models. This is model elaboration. In addition to being a frequently-engaged scientific process, model elaboration is important for learning, because it involves students in making connections across elements of their content knowledge and deepening their understanding of scientific theory. Model elaboration is closely tied to the structure of scientific theories themselves, which can be viewed as populations of models which can be assembled to reason about simple or complex situations in their scope (Giere, 2004). For example, Frederiksen and White's (1988)) module for learning about electricity consisted of a sequence of increasingly elaborated models.

7.1 Rationale, Focal KSAs, and Characteristic Task Features

Stewart and Hafner (1991) identify four ways that engaging in model elaboration benefits students' ability to reason scientifically. These include:

1. *Learning more efficient procedures for generating data.*
2. *Developing within-model conceptual insights.*
3. *Linking models because they share objects, processes, or states.* This involves generalizing from special cases.

© The Author(s) 2017
R.J. Mislevy et al., *Assessing Model-Based Reasoning using Evidence-Centered Design*, SpringerBriefs in Statistics, DOI 10.1007/978-3-319-52246-3_7

4. *Linking models to produce a larger framework.* This entails development of overarching principles that traverse a wide range of problems. For example, Passmore and Stewart (2002) describe a curriculum that includes elaboration of the Darwinian model:

> Once students had initial experiences composing Darwinian explanations and had explicitly considered the components of an appropriate explanation, they were given a data-rich case from which they were expected to develop a more complete Darwinian explanation. This case was designed to provide students with an opportunity to investigate a change in a trait over time, to use the natural selection model to explain that changes, and to support their argument with appropriate pieces of evidence. We intended to create a setting in which it was necessary for students to deepen their understanding of the components of the natural selection model in the course of using those components to create an explanation for the case phenomenon (p. 194).

To develop this quality of knowledge, students must engage in activities that afford this extending and restructuring of their understandings of models. Assessment tasks that address model elaboration will extend a model or address interconnections between or within models. Task situations typically present a situation in which familiar or currently-targeted models are required, but combinations or connections among them are required to formulate a model for the situation. Connections across individual-level and species-level models in evolution and between quantum and classical models in physics illustrate opportunities for learning and for assessing model elaboration.

A simple example of a model elaboration task can be obtained by nesting arithmetic schemas in a multi-step problem. Extending the single-schema task in Fig. 5.1 yields the two-schema situation shown as Fig. 7.1.

Whereas the original problem required forming a reconception of the situation in terms of the Change schema, this two-step problem first requires the formation of a Vary schema to translate kilometers traveled into miles traveled, which as a composite fills in a slot in a Change schema. As a Work Product that emphasizes the relationships among schemas in multi-step problems, the student could be asked to drag the representation of one schema into another, then fill in the slots of the more complex assembled representation with the information given in the problem statement.

A central Characteristic Feature for model elaboration tasks is a situation for which an elaborated model is needed, requiring linkages between models or extensions of the elements of a given model. Note that this characteristic is only fully understood in light of the students' history, in that extending a model in a given context may be a new experience to one student, but to another student involve simply applying a familiar already-synthesized model.

Klaus Frisch recently drove his American-made automobile 265 *kilometers* from San Diego to Santa Barbara to see his parents. When he arrived at his parents' house, he noticed that the odometer of his car registered 45631 *miles*. What was the odometer reading *in miles* before he made the trip? (Hint: 1 kilometer = .6 miles)

Fig. 7.1 A task requiring the nesting of two arithmetic models

If constraining students to model elaboration rather than model revision is desired, the situation or data should be compatible with the models accessible to students. The task solution must involve combining or making additions to existing models. Examples include embedding a model in a larger system, adding more parts to the model, or incorporating additional information about a real-world situation into the schema the model represents that in some way modifies the modeled representation.

7.2   Additional KSAs

Because model elaboration regards the structure of knowledge, Additional KSAs concerning subject-matter knowledge are of particular importance. Content knowledge is a prerequisite to model elaboration. Thus, failure on a model elaboration task can be due to lack of subject-matter knowledge. Only if we can rule out lack of subject-matter knowledge as an explanation for poor performance can we infer troubles with model elaboration. The hint in the two-schema odometer problem provides the relationship between miles and kilometers to remove not knowing this fact as an alternative explanation for poor performance.

As usual, familiarity with task expectations, materials, and procedures are Additional KSAs that enable or hinder performance and must be dealt with by design choices for materials, Work Products, and evaluation procedures in light of the testing population, purpose, and context.

7.3   Variable Task Features

The substance and particular models involved in a model elaboration task are central Variable Features of tasks. Learning tasks can build on models that students are already familiar with in order to maximize opportunities to further students' understanding. The Genetics Toolkit example discussed in Box 3 is such an example.

A model elaboration task can address elaborating or extending a given model, or connecting multiple models. One could split the model elaboration into two more narrowly defined *design patterns* along this distinction.

As with other *design patterns* in this collection, Variable Task Features include whether the task provides the data or situation that is the object of modeling, whether the aspect is the sole focus of the task as opposed to being part of a larger activity, whether the task is to be addressed by an individual or a group, and how much or what kinds of support to provide. One kind of support concerns the model (s) that are the focus of elaboration: Are hints or direct instructions offered for the model(s) to be elaborated, or are they to be generated, unprompted, by the student?

Use of knowledge representations and tools is also a Variable Feature. Involving a representation or tool can support for students who are familiar with it and increase the evidentiary value of the task. But when the assessor does not know whether students are familiar with representations or tools, requiring their use introduces an alternative explanation of poor performance and degrades evidence about Focal KSAs.

The degree and complexity of elaboration required is a Variable Feature. Is a straightforward elaboration of a familiar model required, or less obvious extensions within or across models? Or are multiple models involved?

Box Genetics 3. Model-Based Reasoning Tasks in Genetics: Model Elaboration

It is common for students to first learn a simple model, then learn to extend it to incorporate more variables or additional situations. Model elaboration can be assessed by presenting a student with a familiar model and additional information that requires extensions of the original model to accommodate the new information.

In genetics, students generally learn about the simple dominance model first. They will then be given problems that may ask them to determine the possible outcomes of a cross, or based on the outcomes of a cross, to identify the dominant and the recessive traits. Students may then be given the information that for some traits there are more than two alleles. The Virtual Genetics Lab presents situations in which there are three alleles for the color of a bug. In this case the possible colors are blue, green, the possible alleles are represented as A, B, C, where A is dominant to B and C and will lead to a blue bug. B is dominant to C and will lead to a green bug, and C is recessive to both A and B and will lead to a red bug. In this lab, students are given possible bugs and are asked to cross them in order to determine which traits are recessive.

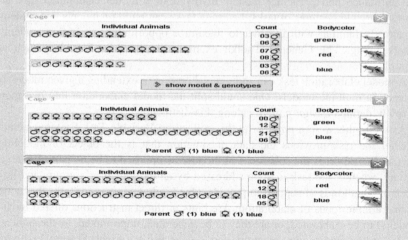

The simple dominance model that the students had previously learned is a model for alleles and rules for their combinations. It fits a number of real-world situations. In this task, students must extend their understanding of the rules in the simple dominance model with two alleles to situations in which there are more than two alleles, combining under an extended set of analogous rules. They must realize that each pair will have a dominant and recessive trait, and while one allele may be dominant in one situation, it can be recessive in another. This elaboration extends their inheritance model to more real-world situations. For this task, familiarity with the simple dominance model is an Additional KSA. In a class where the teacher knows the students are familiar with the simple dominance model, this knowledge can be presumed. In a large-scale test, the test designer could choose to provide the simple dominance model, in order to better focus the evidence on model elaboration. This basic task could be varied with respect to the trait in question, the number of alleles, and the types of relationships in the model. For example, is trait dominance strictly ordered or are there instances of circular dominance? Different choices can increase or decrease the difficulty of the task, while still providing evidence about the students' capabilities with model elaboration. For this problem, the students are only asked to select organisms to cross, and then click a button to obtain the results. Variants of the task could require more complicated procedures for students to test out their model, such as requesting cross of offspring from an initial set of crosses.

Work products could include the students' answers about which alleles are dominant to other alleles, in open-ended or multiple-choice forms; the crosses that they performed; and explanations of their reasoning. With regard to Potential Observations, instructors could characterize how systematic a student's choices of crosses were. From explanations, they could determine if a correct answer was based on appropriate or inappropriate reasoning, through a correct or in some way flawed model.

7.4 Potential Work Products and Potential Observations

From tasks with Characteristic Features of this *design pattern*, students generate Work Products that may include representations of their elaborated models (including, for example, nested representations of model schemas as in Marshall's SPS or STELLA models that incorporate familiar sub-models), oral or written explanations, traces of their steps while developing their elaborated models, and mappings of their elaborated model with a situation.

Potential Observations that can be identified with these Work Product, including the accuracy and completeness of the linkages in elaborated models, the extent to which the elaborated model is accurately linked to the situation, and the quality of explanations for their finished product and how they got there. Potential Observations of particular importance in model elaboration are (1) appropriateness in the region in the model space where the extensions or connections are required, and (2) appropriateness of the correspondence between the modeled situation and the posited model in the region in which the elaboration is required.

7.5 Some Connections with Other Design Patterns

Model elaboration can be considered a special case of model formation, in that the aim is to develop a modeled conception of a situation (then perhaps carrying out further reasoning with it). However the emphasis in model elaboration is on what is happening in the model layer with respect to extensions of models or connections between models, even though these may be motivated by the real-world situation.

Model elaboration is similar to model revision, in that a given model or a set of unconnected models does not account properly for the target situation and reformulation is required. It differs by its more particular focus on extensions and combinations of models rather than modifications within a given model's aegis, to rectify discrepancies in the model/data correspondence.

Model elaboration is also often connected with model evaluation and model revision, when the elaboration is not simple and straightforward: Does an elaboration fit the situation or data? If not, where and why? How might it be revised?

Chapter 8
Model Articulation

Abstract The Model Articulation *design pattern* supports developing tasks to assess articulating meanings between systems associated with a model. Focal KSAs concern making the connections, translations, or re-representations of information within a model system, across representational systems associated with the system, such as diagrams, equations, graphs, and digital or physical models. A Characteristic Feature of such tasks is the need to translate meaning or information across multiple representation systems these forms.

As conceptual knowledge structures containing content, procedural, and strategic information, scientific models admit to representation in a variety of forms. These are indicated by the layers on the highest plane in Fig. 1.1. Examples include force diagrams in physics, Punnett squares in genetics, and algebraic equations. Representations can be quantitative, qualitative, or a combination of these, and there are often multiple representations associated with a scientific model. For instance, both force diagrams and algebraic equations are used to express Newton's laws. In structural equation models, interconnected equations, path diagrams, and computer code link latent and observed variables. These representational systems allow us to reason in different ways about different aspects of a model and real-world situations.

Although these representations vary in their form, they share a symbolic nature– circles, squares, and arrows in structural equations diagrams, for example, and alphanumeric characters and operator symbols in mathematical equations. A symbol system encompasses a set of symbols, interrelationships among symbols, and valid operations for acting on symbols. The markings, notations, or sounds of symbol systems are distinguished from the meanings they denote (Greeno, 1989). A model with one or more such representations can be conceived as a system of objects and entities (e.g., model genes, model particles, model molecules) and the relationships and processes that characterize them (e.g., modeled mutation, modeled atomic structure). The relationship between the qualitative entities and relationships of the model or abstract systems establishes the meaning of symbols and operations in the symbol system.

R.J. Mislevy et al., *Assessing Model-Based Reasoning using Evidence-Centered Design*, SpringerBriefs in Statistics, DOI 10.1007/978-3-319-52246-3_8

8.1 Rationale, Focal KSAs, and Characteristic Task Features

The importance of the ability to navigate from one representational system to another is illustrated in findings such as Larkin's (1983) study of physicists' reasoning. When presented with a force problem, experts first took a qualitative approach, identifying salient entities and relationships and singling out appropriate models for solving the problem ("This is an equilibrium problem"). From their resulting understanding at this qualitative level, they proceeded to build a set of equations (using the associated symbol system) that corresponded to the situation, then solved the problem by working through the equations. By connecting the physical situation to a description in the semantic terms of the model and in the symbol system in turn, the work within the symbol system acquired a situated meaning in the problem at hand.

As Greeno (1989) notes, however, much learning in classrooms targets students' ability to reason within a particular (typically symbolic) representational system— fluency with the symbolic notations, operations and relationships of linear algebra, for example. While these skills are necessary for reasoning with scientific models, just being able to carry out manipulations strictly within a symbol system layer is not sufficient. Ability to reason between representational systems is an essential aspect of inquiry and is thus an important target for instruction and for assessment. This includes translating meanings between the semantic system of a model and an associated symbolic system, or from one symbolic system to another. For instance, evidence of students' ability to translate force diagrams into mathematical equations and vice versa supports claims about their ability to reason with models for force and motion.

The Model Articulation *design pattern* supports developing tasks to assess articulating meanings between systems associated with a model. Focal KSAs concern making the connections, translations, or re-representations of information across representational systems associated with a model system. This includes the mappings between the semantic entities, relationships, and processes within the model (the middle layer of Fig. 1.1) and formal representations (the upper layer). It also includes expression between widely-applicable representational systems such as graphs and mathematical expressions, as contextualized within a given scientific model.

Characteristic Features of tasks that assess model articulation are the involvement of multiple representation systems and the need to translate meaning or information across these forms. This may take various forms, such as relating semantic and mathematical formulations, or semantic formulation and physical models, or two symbolic representations such as equations and graphs within a context and using the model of interest.

Model articulation differs from model formation in that it focuses on reasoning at the semantic and associated representational layers above it rather than the correspondence between a model and a real-world situation. However, when model articulation is addressed within the context of a real-world situation, the situation imposes constraints on connections among representations and provides situated meanings for connections among representations. Interpreting these is an aspect of model articulation.

8.2   Additional KSAs

As usual, Additional KSAs are content knowledge and familiarity with task expectations, materials, and procedures. Inferences made about model articulation would ideally be able to assume that students are already able to reason *within* the system at issue, since failing to do so is an alternative explanation for task failure. Mapping between force diagrams and algebraic representation in mechanics, for example, can fail if a student is insufficiently skilled with the calculus needed to express a targeted relationship. Therefore, while the Focal KSA in this *design pattern* targets ability to articulate *between* systems, knowledge *within* systems serves as an Additional KSA.

8.3   Variable Task Features

Which model system is addressed is as always a Variable Feature of tasks. Within this selection, designing Model Articulation tasks also presents the choice of which representations to address. A key distinction is whether the targeted articulation is between the semantic layer of a model and a representational system, or among representational systems.

Tasks motivated by this *design pattern* can vary in the number and combinations of systems included. Some tasks may include only a single symbol system and a single model system, and ask students to describe the symbols in terms of the model entities, or vice versa. Alternatively, tasks may require students to consider two symbol systems associated with a model and re-express the meaning of an expression in the one system with an expression in the other. Potential Observations here could include the accuracy and completeness of the mapping between systems.

Other tasks may ask students to express a prediction in terms of one system based on a given representation in a second system. "What will happen to the velocity of the ball described by equation b?" calls for articulation between the mathematical representation and the semantics of the Newtonian model. Physical representations can also be used, as when elementary students are asked to express a subtraction problem with physical objects (e.g., $6-2 = ?$ can be represented as removing two blocks from a pile of six).

The complexity of the systems and mappings are variable as well. Requiring transformations within systems as well as across systems increases task difficulty. (It also increases the requirements for the Additional KSAs for capabilities with those systems.) Such tasks require deeper understanding of the relationships among the components of the model system. This may be construct-relevant in some contexts, and therefore appropriate as is, but it may be preferable to scaffold within-system operations in order to sharpen the focus on the articulation between systems.

Another central Variable Feature is whether the articulation in focus is prompted. On one hand, a task designer can explicitly call for a mapping or interpretation between the semantic and a specified symbolic system, or between two specified symbol systems. On the other hand, evidence for articulation between systems may be sought without prompting in an open-ended solution to a problem or during the course of an investigation. The resulting Work Products may or may not contain evidence; if they don't, it will not be possible to evaluate the relevant Observable Variables. But unprompted evaluation of model articulation is necessary when a task is meant to assess the student's recognition of the need and appropriateness of alternative expressions to support certain reasoning within the model system.

8.4 Potential Work Products and Potential Observations

The main Work Products that convey evidence about model articulation are re-expressions of information about elements or relationships across multiple representations, and explanations of such representations. These could be in closed form, as with multiple–choice alternatives that offer different re-expressions or explanations; they could be constructions of representations either from scratch or from partially-completed forms; or a trace of activities leading to articulation across representations. As noted, an open-ended Work Product may either be prompted ("show the mapping across these two representations") or unprompted. When prompted, the student will be asked to construct or complete a second representation given information in terms of a first representation. When not prompted, the Work Product is the trace, the final or intermediate products, and/or a solution protocol from an open-ended solution—in which the student may or may not have provided the desired evidence. The directive "show your work" helps ensure that the representations will be provided, as long as the student is familiar with the task format and expectations.

Potential Observations address correctness and quality of the required mappings or explanations of symbolic expressions. Multiple aspects may be evaluated. When the Work Product includes an explanation, Potential Observations include the quality of the explanation of what is common across the systems and what differs, and how it matters for reasoning.

8.5 Some Connections with Other Design Patterns

Model articulation will often be pertinent in multiple-step tasks, after model formation. There are several reasons. First, proficient use of symbol-system representations is generally preceded by the formation of a model in the semantic layer—that is, in terms of the model entities, relationships, and processes (Larkin, 1983). These are the connection to the symbol system, rather than the symbol system representation being mapped directly to the situation. Any reasoning that uses a symbol-system representation has involved model articulation.

Second, solving a problem often requires transforming information about a real-world situation from expression in one system to another for a different purpose. A table may be a good way to represent the outcome of an experiment, but this representation is not as suited as an algebraic expression or statistical graphic for quantitative manipulations. The genetics example shown in Box 4 illustrates this point. The results of crosses are shown as tallies, but they must be transformed into a Punnett Square in order to employ the logical and statistical machinery associated with this standard representation for reasoning about crossing results.

Third, when the results of symbol-system operations are completed or when the outcome of an investigation is summarized, the outcomes must be expressed in a form that connects the outcomes with the model's semantics. Articulation to a representational form that is tuned to communication is required, which is necessarily not the same as a form that is tuned to the operations. Labeled path diagrams are used to report the results of structural equations modeling, for example, while matrix algebra and computer code were the representation through which estimation is carried out.

Box Genetics-4. Model-Based Reasoning Tasks in Genetics: Model Articulation

An instructor interested in determining how well students understand a given model can use a Model Articulation task to see if they can reason across representations. In this genetics example, students are presented with the Virtual Genetics Lab representation of a cross as shown below. Students may then be presented the following tasks:

(1) Use a Punnett Square to explain how the results of the cross were obtained.
(2) What is the expected percentage of offspring with a short body type? How did you obtain that answer?

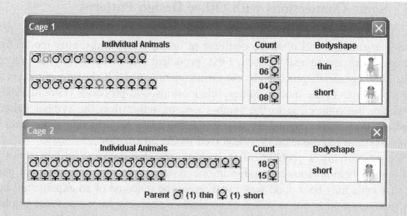

For this problem students must be able to articulate how a cross is performed, and must understand the relationship between the results given for
cage 2 and the entries of a Punnett Square. They must then be able to move
from a graphical representation to a numerical representation in terms of the
percentages associated with the possible body types.

As with all model articulation tasks, this type of problem requires students
to be familiar with multiple representations of the subject matter. The focal
KSA is the transition between different representations. For this problem
three different representations are given; an instructor could remove one or
add more to decrease or increase difficulty. The difficulty will be affected by
how familiar students are to each of the representations.

Work products would include the Punnett square produced by the student
and the student's explanations. From these work products an instructor could
determine how well students understand the concept of crossing and how well
they are able to use multiple representations to obtain conclusions about the
results of the given cross.

Chapter 9
Model Evaluation

Abstract Model evaluation is examining the appropriateness of a model for a given situation or data. The Focal KSAs in model evaluation tasks are the capabilities to determine whether, how well, or in what aspects, a model is appropriate for a given situation. Potential Observations may include whether students identify cues of model misfit; whether particular areas, patterns, or unaccounted-for features of the situation are identified; and whether hypotheses for the model-data discrepancy can be proposed. The Model Evaluation *design pattern* is tied closely with the Model Use and the Model Revision *design patterns*.

Model evaluation is examining the appropriateness of a model for a situation or data. This may be as straightforward as answering the binary question of whether or not the model fits the data, it may require an explanation, or it may involve an investigation of how well or in what respects the model fits and fails. While tasks can be devised that focus primarily on model evaluation, this aspect of model-based reasoning is intimately connected with several other aspects of model-based reasoning. Model evaluation is tied inextricably with model use. In order to evaluate a model, students must be able to reason through the model to project its facsimile of the salient features of the situation, whether qualitative or quantitative, because comparing these projections with the actual situation is the basis of model evaluation. While it may be hard to separate model use and model evaluation (and often unnecessary), tasks can be designed to focus on model evaluation for more targeted instruction and assessment.

9.1 Rationale, Focal KSAs, and Characteristic Task Features

In any type of model-based reasoning, students need to be able to connect the real-world situation and the model (the arrow in the lower left corner of Fig. 2.1). Without model evaluation, students have no justification for why one model may be

© The Author(s) 2017
R.J. Mislevy et al., *Assessing Model-Based Reasoning using Evidence-Centered Design*, SpringerBriefs in Statistics, DOI 10.1007/978-3-319-52246-3_9

better than another, and therefore may not be able to determine an appropriate model. In real-world situations where the model is not provided, students will have difficulty addressing the problem if they cannot evaluate, as well as propose, candidate models.

There are three basic ways to elicit evidence about model evaluation in tasks: a model is provided and the student must address its appropriateness; multiple candidates are provided or suggested and the student must determine their suitability; and a model is not given and the student must formulate a model. The first two focus attention on model evaluation specifically, while the third integrates model evaluation into the flow of inquiry. In an investigation task, a rubric can include assessing model evaluation as it occurs in students' ongoing procedures or in their final presentations.

Model evaluation is often prerequisite to model revision and model elaboration; it is necessary to determine how and how well a model fits a situation before one can improve it. The quantum revolution was motivated in part by Newtonian and field mechanics' failure to account for the photoelectric effect and "block box" radiation. Tasks designed to provide evidence about model evaluation can be extended by model revision or elaboration.

A classic example of model evaluation in statistics is multiple regression. A variety of model-checking tools are used to examine how well the model fits and the structure of the relationship between the independent and dependent variables (e.g., Belsley, Kuh, & Welch, 1980). Mosteller and Tukey (1977) show how to use them in inquiry cycles. A simple regression model posits a linear relationship between the two variables, or $y = ax + b$. An analyst may hypothesize that age and strength are related, such that as people grow older they get stronger. To evaluate this linear model, would study the fit of the regression model to data on subjects' ages and strength. One method is to test the theory graphically (note the articulation needed between an equation and graphical representations). Does the pattern in the data points look like what one would expect under the theorized relationship? For age and strength, the researcher may find that the graph looks more curvilinear, or is linear only within ranges, thus moving toward to model elaboration or model revision. Figure 9.1 shows situations where the linear regression model appears suitable, there is a curvilinear relationship that it cannot capture, and an outlier that renders the regression line misleading.

Baxter, Elder, and Glaser's Mystery Boxes (1996) combines model evaluation, model use, and model revision is. Students are presented six different boxes with some combination of elements among a light bulb, wire, and batteries, and they must perform tests to determine what is inside each box (Fig. 9.2).

The students have been studying a model for simple circuits with these components. In this hands-on task, they must use their understanding of this model to determine what sub-model fits each of the boxes. They must determine which tests (connecting the terminals of the mystery box with just a wire, with a battery, with a light bulb, and so on) are appropriate in narrowing down the choices for the submodel. Interpreting the results of a test requires reasoning through each provisional model to predict what would be observed if it were the true configuration (model use), then determining whether the observed result is consistent with the

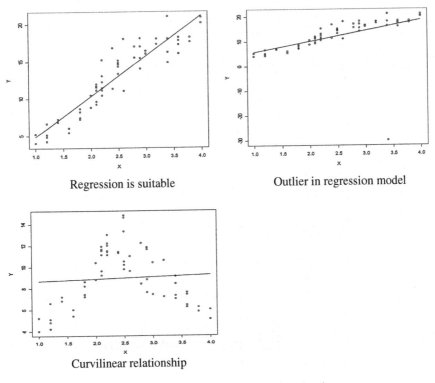

Regression is suitable   Outlier in regression model

Curvilinear relationship

Fig. 9.1 Examples of a simple regression model with three data situations

prediction. Comparing this prediction and what actually happens is model evaluation. Generally a single test is not sufficient to determine conclusively which configuration is inside a box, so the results of multiple tests must be synthesized to evaluate each possibility. This feature of the task leads to some Potential Observations concerning evaluation strategies, which will be discussed in Sect. 9.4.

The Focal KSAs in model evaluation tasks are the capabilities to determine whether, how well, or in what aspects, a model is appropriate for a given situation, be it a real-world scenario or already-synthesized data. This can include identifying relevant features of the data and the model(s) under investigation, and evaluating the degree and nature of the correspondence between them. In the examples, students must be able to examine the data given (the regression data set or the physical mystery boxes) and determine model fit and suitability.

Characteristic Features of tasks designed to assess model evaluation include a target situation and one or more models. The models should be able to be examined in light of the situation and the data. In the regression example, the situation is predicting an outcome variable, and the model is the statistical relationship between the variables. In Mystery Boxes, the situation is determining what circuit in a given box produces the observations the tests produce. The family of models at issue is the set of completed circuits that can be formed from the configurations of elements within the boxes and elements that connect the terminals.

Find out what is in the six Mystery boxes A, B, C, D, E, and F. They have five different things inside, shown below. Two of the boxes have the same thing. All of the others have something different inside.

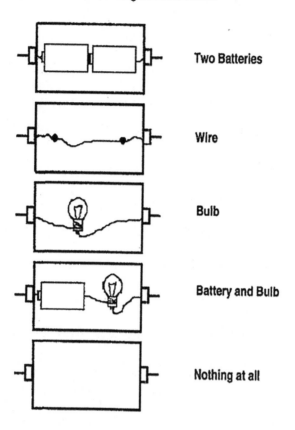

For each box, connect it in a circuit to help you figure out what is inside. You can use your bulbs, batteries, and wires any way you like.

Fig. 9.2 Mystery box task (Baxter et al., 1996)

Box Robotics-4. Model Evaluation

Model Evaluation is central to the robotics task. As in many engineering problems, theory and experience guide the design of an artifact that will produce desired results under given constraints (Simon, 1996), but solutions may require repeated trials and successive approximations. In this task, Model Evaluation requires analyzing, critiquing, and diagnosing the behavior of a simulated or physical rover in each trial. That behavior arises from the configuration being tested in that trial. It is the basis for revising the model (discussed in the following section). To know how to revise the model (simulated or physical), one must understand how the observed behavior departs from the desired behavior, and reason through the model to determine what characteristics of the artifact produced less-than-optimal results. In this task, this aspect of Model Use is thus embedded in Model Evaluation in every testing cycle. In the simulation phase of the investigation, the student can view the rover's behavior and additionally ask to view generated graphs as in Fig. R4. Is the rover traveling up the ramp? At what rate of speed? Does it stop at some point along the way? Are its wheels just spinning?

The particular Focal KSA in evaluating this engineering model is characterizing the artifact's behavior in the criterion situation, with particular attention to its correspondence to the desired behavior. The Characteristic Feature in each testing cycle is behavior through the model in comparison with the desired behavior. This is so in both the simulation or physical phases. It is the necessary feature of a situation to evoke Model Evaluation. Note that this aspect of model-based reasoning is not isolated as an encapsulated task. Rather, each instance is marked by the student working herself into this situation—perhaps without even recognizing it, and thus failing to carry out the process. Further, because it is a part of each testing cycle, one student might work herself into only one such situation, while another works himself into five of them—all evidence-bearing opportunities for assessing the student with respect to this aspect of model-based reasoning.

Again a critical Variable Task Feature is whether the modeling is carried out in the simulation space or the physical space. (Another potential value of this variable task feature would be to carry out the initial design work with paper and pencil renderings and approximations of behavior through equations.) Both phases entail the Additional KSA of domain knowledge with respect to the electric circuit, motor functioning, and gearing ratios and their implications. However the simulation/physical design variable brings with it a cluster of other Additional KSA demands, associated Variable Task Features, and Work Products. The Additional KSAs concern proficiencies for working in the appropriate space, either simulation tools and affordances or the capabilities to run, observe, and record behavior of a physical rover. (It might be mentioned that one could attach an accelerometer to the physical rover, and obtain more information—and at the same time engage corresponding Additional KSAs.)

Another Variable Task Feature that differs across phases is the amount of scaffolding. Built into the simulation environment is a tool call the Learning Companion—a kind of coach that supports Model Evaluation and the upcoming Model Revision aspect and the overarching Model-Based Inquiry activities. As noted previously, the log file of a student's activity in the simulation space is captured, and is used to keep track of a student's rover design in each trial and its behavior. It counts the attempts, and offers feedback that is likely to be helpful. Figure R5 shows its logic.

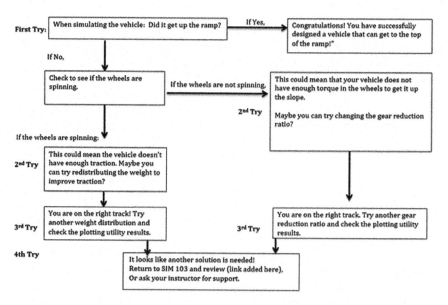

Fig. R5 Learning companion feedback for vehicle trying to climb a ramp

The physical phase of the task does not engage the Learning Companion. The intent is that the support it affords in the simulation phase will have provided the student with a schema of kinds of rover behaviors to look at and what they mean (and later, what to do next).

9.2 Additional KSAs

In addition to the Focal KSAs described above, assessments of model evaluation can require different levels of domain knowledge and familiarity with the type of task or model being evaluated. With regard to domain knowledge, being able to evaluate the fit of the model depends on being able to identify mismatches between a model and a situation. The more subtle the mismatch and the more it depends on the particulars of the model, the more critical the Additional KSA of domain knowledge becomes; domain knowledge sets expectations about what features are relevant and what relevant patterns should look like. Thus an assessment meant to focus on the process of model evaluation per se would use familiar models and situations. An assessment meant to address model revision jointly with knowledge of particular models can validly have a high demand for the substantive aspects of those models.

Additional KSAs also include familiarity with the methods used to evaluate the model, and the standards and expectations in the field. As noted below regarding Variable Task Features, a designer can use scaffolding to reduce the demand on Additional KSAs consisting of background knowledge, planning, and evaluation methods such as graphical and statistical tools in the regression task. As always there are tradeoffs: requiring fit indices to be calculated by hand increase demands for computational procedures, but requiring the use of a particular computer program instead brings in Additional KSAs for using the program.

In the Mystery Boxes task, students' knowledge of circuits materially affects the difficulty of the task. The students in Baxter's study had just completed a unit on circuits, so the evidentiary focus of the task for them was in planning and carrying out the testing procedures to infer what was the boxes. Students who are not familiar with circuits might not be able to reason through possible combinations of elements (model use) to carry out model evaluation (although the tasks could be instructional activities to help them learn about circuits). Students also need some knowledge to connect the components. These Additional KSAs concerning procedures would be circumvented in a simulation version of the task. Further, students could be scaffolded in steps of evaluating the boxes in order to reduce demands on planning and organizational capabilities. Baxter et al. chose not to, because evaluating how the students planned and explained their procedures was central to their research.

9.3 Variable Task Features

Model evaluation tasks can vary as to the type and complexity of the model(s) to be examined. Model evaluation can be prompted or unprompted (implicit) in a given task. That is, tasks can present models to students and explicitly direct the student to examine them, alternatively, students have to determine the models themselves, and indeed whether or not to evaluate fit. Whether the model at issue fits, and if not, the degree and nature of misfit, can also be varied. The type and complexity of the model evaluation methods may differ. These choices can be used to increase or decrease demands for particular aspects of the Focal KSAs and Additional KSAs. Different choices can provide more or less information, often trading off against convenience and economy of scoring procedures. More open-ended tasks take longer for students to complete and present more challenges for scoring, but provide more evidence about students' reasoning, and incorporate model evaluation into the inquiry process.

In regression tasks, for example, the number of predictors can be varied. The students can also just be asked to use graphical displays to explain why they believe the model fits or does not fit, or to use statistical methods or graphical methods to justify their conclusions.

In Mystery Boxes, the medium of the task could vary: physical boxes, an interactive computer simulations, or static paper-and-pencil representations. The last is simplest and easiest to score, but it places a greater demand on model use for projecting the results of different configurations, as the task environment itself no longer provides feedback. The number and contents of the boxes could be modified, as well as the information students have about what they might contain. All of these modifications can affect task difficulty and raise or lower demands on different aspects of knowledge, both focal and additional.

Mystery Boxes illustrates design choices about the kind and amount of scaffolding to provide. As mentioned, Baxter et al. did not scaffold the process so they could obtain evidence about students' planning and self-monitoring. A different scaffold would be a chart of the results of tests when applied to different configurations. It removes most of the demand for reasoning through the circuit model and shifts the focus to model-evaluation procedures. Whether to do so depends on the intended examinees and the purpose of the task.

9.4 Potential Work Products and Potential Observations

The simplest Work Product for a model evaluation task is the indication (in whatever format specified) of whether or not the model fits or which model fits. It is also least informative. The next more informative option is having the student provide qualitative and/or quantitative indications of degree and nature of fit and misfit. Statistical tests, graphs, or other representational forms for model evaluation can be evoked as Work Products. These "final product" Work Products can be accompanied by an explanation (verbal or written) of why and how the student reached the conclusion. This can include verbal or written explanations of the hypotheses formulated (regarding fit) and the methods used to test them, including the output from model-fitting tools. Written, verbal, or computer-tracked traces of the actions a student performed can also be captured as Work Products.

Compared with simple choice Work Products, explanations are particularly useful in determining if a student understands the situations and models well enough to evaluate them critically. The formality of an assessment may dictate the format required as well as the depth expected. Solution traces can take various forms, such as computer logs of actions students take through a computer interface in a simulation investigation, a video recording of actual performance, or a written trace by the student of the steps in their evaluation. All of these examples provide more evidence about the efficiency of the student's model evaluation procedures than a final solution. Baxter et al. (1996) found the last of these particularly useful for evaluating the rationale behind students' decisions.

Potential Observations in model evaluation tasks can thus address the comprehensiveness and the appropriateness of the hypothesis generated through the model for evaluation, the appropriateness of the evaluation method(s) used to assess model fit, the efficiency and the adequacy of procedures the student selects, and the correctness and thoroughness of the evaluation. This can include whether students identify cues of model misfit; whether they identify particular areas, patterns, or unaccounted-for features of the modeled situation; and whether they propose hypotheses for model-situation discrepancies. All these Potential Observations provide evidence about model evaluation in context, but all also require some degree of understanding of the substance of the situation, as Additional KSAs.

More complex Work Products provide the opportunity to explore these qualities more deeply. Simpler Work Products—such as selection of a best-fitting model—provide less information but, on the other hand, allow the aspect of proficiency to be targeted more precisely. The quality of the explanations given, as well at the quality of the determination of how an ill-fitting model might affect inferences resulting from that model, also can be examined. How well a student integrates results from multiple methods of testing can be observed, along with how well a student is able to indicate which aspects of the model and data do not fit.

In the regression example, the Work Products can include the output from a formal model fitting tool and an explanation of the conclusions drawn from observing this output. From the output of graphical and statistical model-fitting tools, an assessor can evaluate the quality of the explanation and the appropriateness of the tools used and how they were applied.

In Mystery Boxes, Baxter et al. gathered as Work Products the students' initial plans, their strategies, and explanations of their solutions, and traces of their activities in the form of written logs and think-aloud protocols. The researchers evaluated the "explanation" Work Products for what students expected if a certain combination of components was inside the box—the aspect of model use that is integral to evaluating a proposed model. From the trace of students' activities, the investigators observed and evaluated how flexible the students were as they monitored their results. The Work Products made it possible to create observations that addressed not only the end results (the students belief about the contents of each box) but also how well the students were able to interpret the results of each of the tests to determine which further tests, if any, were needed.

9.5   Some Connections with Other Design Patterns

The Model Evaluation *design pattern* links closely with model use. It can even be difficult to develop tasks that assess only model evaluation. However, tasks can be designed to emphasize either model use, model evaluation, or both. This may be accomplished by scaffolding whichever aspects of reasoning (if any) are not the intended focus of the task (e.g., providing a table of test results in the Mystery Box tasks).

The Model Evaluation *design pattern* also is associated with model revision and model elaboration because in order to determine if a model needs to be revised or elaborated, some model evaluation usually needs to have been performed. Model revision tasks can be designed to minimize model evaluation by presenting the students with a situation and a model they are told is inadequate in a given way, which they need to revise or elaborate.

Chapter 10
Model Revision

Abstract Model revision is modifying a given model in a given situation, so that its elements better match the features of the situation for the purpose at hand. Model revision tasks feature a situation to be modeled, a provisional model that is inadequate in some way, and the opportunity to revise the model in a way that improves the fit. Provisional models may be provided or arise through the students' work, possibly multiple times, in more encompassing tasks. Work Products can include the choice or the construction of a representation of the revised model, and an indication of the problem with the initial model and how modifications could address the issue.

Model revision is the aspect of model-based reasoning that allows us to speak of the inquiry *cycle* rather than the inquiry *sequence*. We form a model for a situation and purpose, we reason through it to evaluate its aptness—and more often than not, find it isn't quite right. We must then use the clues about where and how the model doesn't fit to modify it and improve the correspondence to better serve our purposes.

10.1 Rationale, Focal KSAs, and Characteristic Task Features

The Focal KSA in model revision is the capability, in a given situation, to modify a given model so that its features better match the features of that situation for the purpose at hand. This capability can be further differentiated into recognizing the need to revise a provisional model, modifying it appropriately and efficiently, and justifying the revisions in terms of the inadequacies of the provisional model and the intended use.

Model revision tasks feature a situation to be modeled, a provisional model that is inadequate in some way, and the opportunity to revise the model in a way that improves the fit. Box Genetics-5 is an example from the Virtual Genetics Lab that requires model revision. The examinee is presented with a situation and a model that has been proposed by a hypothetical student. The examinee must evaluate and then revise the proposed model.

© The Author(s) 2017
R.J. Mislevy et al., *Assessing Model-Based Reasoning using Evidence-Centered Design*, SpringerBriefs in Statistics, DOI 10.1007/978-3-319-52246-3_10

The Biomass project (Steinberg et al., 2003) suggested an adaptive procedure to elicit evidence about model revision. A student would be provided the results of a first crossing of animals with an unknown heredity structure for a given trait. These results would be consistent with two different inheritance structures. The student would be asked to propose a plausible model for the dominance structure among the alleles. The results of a second crossing would then be presented that contained information to distinguish between the models that fit the first crossing — and the second-crossing results presented to a student would be selected to be inconsistent with her first response and consistent with the one that she did not propose. Model revision is required no matter which model the student first hypothesized.

Box Genetics-5. Model-Based Reasoning Tasks in Genetics: Model Revision

A task schema that can be used to assess Model Evaluation and Model Revision is to present an examinee with an incorrect model proposed by a fictitious student. The ways in which the provisional model is incorrect are chosen to highlight whatever features of the substantive model or the evaluation techniques are the target of inference. The task illustrated here was developed by the authors of this paper, but uses a representation from the Virtual Genetics Lab to illustrate the approach.

The background for this task explains that a student Jose has found six bugs in a shed, four with long wings and two with short wings. He decides to investigate the mode of inheritance of wing type. He hypothesizes that there are two alleles and the mode of inheritance is simple dominance. He crosses two long-winged bugs. To his surprise, he obtains the following offspring:

Cage 3			✕
Individual Animals		**Count**	**Wings**
♂♂♂♂♂♂♂♂♀♀♀♀		08♂ 04♀	short
♂♂♂♂♂♂♂♂♂♂♀♀♀		10♂ 03♀	four
♂♂♂♂♂♂♂♂♀♀♀♀♀♀♀		07♂ 07♀	long

Parent ♂ (1) long ♀ (1) long

Model Evaluation is first required. The appearance of a third wing type, four wings, contradicts Joe's hypothesis. Further investigation will show that while there are in fact two alleles, the mode of inheritance is co-dominance: when the two different alleles are combined they produce a third variation of the trait. The data shown above are not conclusive, so repeated cycles involving model formation, crossing, and model evaluation will be required.

Box Robotics-5. Model Revision

Model Revision is central to the robotics task. The student has a provisional model—a given version of a simulated or physical rover—that is based on knowledge of the underlying scientific principles (circuits, motors, gears) and performance of previous versions. It is run in the target environment, and its performance is observed and analyzed (Model Evaluation). If performance is not yet satisfactory, then comes the defining question of Model Revision: How might the model be changed, to more closely match the desired behavior?

Note that Model Use is embedded in answering this question. The student must reason through the existing model to understand why it produced the behavior observed in this trail, and reason through models that differ in selected ways as to how they would behave differently. Focusing on the gear ratio in the rover task, a crucial observation is whether the wheels are turning when it stops making progress up the ramp. If they are not, it is likely that there is not enough torque, and a higher gear ratio will be needed. Carrying out this reasoning is the instantiation of the Focal KSA of Model Revision elicited in this task. Understanding the gear ratio model, the principle of torque, and the behavior of systems with insufficient torque are Additional KSAs. The MOOC supports these with both previous and "just in time" instructional material.

Recall that the task is designed so that the first attempt in the simulation phase will fail, due precisely to not having enough torque. This is a forced instantiation of the Characteristic Feature of the task, "A situation to be modeled, a provisional model that is inadequate in some way, and the opportunity to revise the model in a way that improves the fit." This ensures that all students will encounter at least one occasion to engage in model revision. Some may have more than one, depending on how many inquiry cycles they need to produce a rover that climbs the ramp.

Another necessary Additional KSA required in the simulation phase is understanding the graphical representations (Fig. R4) [refer back to the figure in Box Robotics-3]. These graphs provide information critical to revising the model in a favorable direction: The comparison of wheel-rotation to forward progress over time in a given trial, and the comparison of behavior in the current trial to previous trials. The student is provided this information explicitly in the simulation phase, and can associate it with the directly observed behavior of the simulated rover. The intention is that this scaffolded reasoning in the simulation phase will help students mentally critique the behavior of rovers in the physical space on their own.

The log file of student actions is available as a Work Product in the simulation phase. If the rover has stalled on a given attempt and the student revises the model, the information is used to calculate critical Observables: Do the revisions address the correct aspect of the observed problem? For example, changing the battery is irrelevant; changing the gear ratio is the right aspect of the system to address. If the correct aspect is addressed, is it changed in the right direction? Changing to a higher ratio indicates an understanding of torque implications, even if it is changed too much in this direction. Changing to a lower ratio suggests a misconception, or at least lack of understanding, of the underlying gear and torque scientific model. The Learning Companion (Fig. R5) uses this information to provide feedback in real time, a formative use of assessment. The information could also be used as evidence in a psychometric model to provide higher-level feedback to teachers on their students' understanding (Mislevy et al., 2014).

Work in the physical modeling phase again has less scaffolding and fewer built-in tools. It has an added Model Revision challenge as well: Some rover configurations that can climb the ramp in the simulation space cannot climb it when so constructed in the physical space. (The students can use laser cutters to design their own custom wheels, which adds engagement. But some designs they can make don't have enough traction to make it up the ramp with the same gear ratio and weight distribution that would work with the default wheels in the simulations.) In modeling engineering—in engineers' actual work as in educational tasks—the scientific and simulation models that are available cannot encompass all relevant features of the real-world situation of interest. A Variable Task Feature of model revision in engineering tasks is thus whether the underlying models, whether provided or built by the student, are fully sufficient to design an artifact.

The Work products that are generated in the physical phase of the robotics task include the students' successive versions of the rover. Comparisons among them, in conjunction with each trial's results and the nature of the revisions, are a basis for Potential Observables including the number of attempts (efficiency in Model Revision), the appropriateness of the revisions (revisions in line with the underlying models, such as for the gear box), and indications of misconceptions or foundering. Additional work products that could be captured include videos of the performance, students' open-ended explanations of the rationale of their work, and answers to focused prompts about the scientific models intended to underlie the work.

10.2   Additional KSAs

As with model evaluation, the design of model revision tasks requires choices about the domain knowledge that is involved. Especially in advanced tasks, understanding the scientific phenomenon is increasingly important in detecting anomalies and inadequacies because it sets up expectations for key entities and relationships. The same knowledge is needed to revise the model to remediate its inadequacies: What patterns are not being modeled appropriately, and how might they be modeled? The section on Variable Task Features discusses approaches for addressing the relationship between domain knowledge and model revision.

As usual, familiarity with task type, tools and representations, and expectations for performance are Additional KSAs. Variable Task Features should be controlled in a way that is appropriate to the context and purpose of assessment to remove such alternative explanations for poor performance.

10.3 Variable Task Features

Model revision tasks can vary as to model and substance, and students' familiarity with task format, tools, representations, and expectations. These features can be chosen to manage demands for Additional KSAs including domain knowledge and others that are construct-irrelevant for the targeted inferences. The following techniques address aspects of familiarity.

One way to minimize the demand for domain knowledge in large-scale tests is to make the context simple and familiar, as from everyday experience. While removing the sometimes-irrelevant confound of domain knowledge from model-based reasoning, this approach also removes the sometimes-relevant interplay of domain understanding and model-based reasoning.

An alternative is to craft a task that is based on a more substantive scientific model with which the students are known to be familiar. A desired connection between domain knowledge and model-based reasoning can now be exploited, as the evidentiary focus is on reasoning, given the required domain knowledge. This approach is consistent with the view that to understand a scientific model necessarily includes being able to reason with it. Carrying out this assessment approach requires knowing the student is sufficiently familiar with the model area at issue. It is natural to implement this approach when connected with instructional programs or determined locally by teachers who know what students have been studying.

When one desires in large-scale testing to employ model revision tasks with substantial demands for domain and model knowledge (as in the NGSS performance expectations), evidence about the domain knowledge and reasoning are again confounded. To disentangle them, a task can include multiple directives, some of which address domain knowledge and others of which provide domain knowledge in the course of model revision.

A set of related Variable Task Features in model revision tasks are whether a provisional model is provided, inadequacies of the provisional model are provided, model revision is prompted, and the task situation is interactive. A task that focuses exclusively on model revision provides a provisional model, points out its inadequacies, directs the student to revise it accordingly, and requires no iteration. Sometimes this specificity is desired, to focus attention on revising a particular model during instruction or to obtain evidence about a particular educational objective. This specificity trades off against the natural application of model revision in conjunction with model evaluation, and with model formation and model use more broadly. At the other extreme is seeking evidence in the course of a broader investigation, with sufficiently rich work products to reveal evidence about model revision *if it occurs*, and rubrics to evaluate its quality in terms of Observable Variables. Between the extremes are structured tasks that support the student working through the phases of an investigation, including this model revision (e.g., White & Frederiksen, 1998).

As with the other model-based reasoning *design patterns*, model revision has as Variable Task Features the substantive content, type, and complexity of the model at issue, and the representations and tools that are involved.

10.4   Potential Work Products and Potential Observations

Work Products for model revision tasks can include the choice or the construction of a representation of a revised model, and an indication of the problem with the initial model and how modifications could address the issue. Explanations of how the model was revised in response to the ways in which it was found inadequate can also be required, again as choices or constructions.

If model revision is not prompted, as in unstructured investigations, a more comprehensive Work Product is required; a solution trace, intermediate products, or explanation of steps taken, so evidence will be available as to whether model revision was carried out, and if so, how and with what results.

To produce values of Observable Variables from performances to specific tasks, these Work Products can be evaluated for the appropriateness of the methods used and the modifications made. The quality of the basis for determining that the new model is an improvement also can be evaluated, focusing on the degree to which the inadequacies of the original model have been addressed. A multiple-choice task to this end could offer possible corrections and rationales of varying qualities, while an open-ended task would solicit a student's rationale then evaluate its quality with a rubric.

10.5   Some Connections with Other Design Patterns

Because model revision is central to inquiry, it is worth having a *design pattern* to focus on it. Under what conditions can we get evidence about students revising models, so we can build tasks with these features and so we can recognize those situations within more complicated activities? What ways we can capture evidence about students' thinking about how and why to modify models, and what aspects of their work should we call out for evaluation? Yet because of its very centrality, model revision is difficult to assess in isolation. Model revision is prompted by model evaluation, as we must first decide that a provisional model is in some way inadequate. We then use model formation to propose modifications that better address the situation at hand. We use the revised model to reason forward to its implications for observations that we hope will be in better accord with the situation, and use model evaluation again to determine whether this is so. Perhaps better than any other aspect of model-based reasoning, the Model Revision *design pattern* calls to our attention that these *design pattern*s correspond to distinguishable aspects of activity rather than distinct psychological abilities.

Chapter 11
Model-Based Inquiry

Abstract Model-based inquiry highlights the metacognitive aspects of managing and moving effectively through cycles of inquiry. The Focal KSAs in this *design pattern* are students' capabilities to manage their reasoning across in inquiry cycles. A key Variable Task Feature to consider is the degree of scaffolding to provide students as they move from one aspect of an inquiry to another. All the considerations, design choices, work products, and observations addressed in the preceding design patterns can be involved in a model-based inquiry task.

Distinguishing aspects of reasoning is useful in instruction and assessment, but it is their coordinated use that marks model-based reasoning in practice. We would like to help students learn to move among these aspects of reasoning, often without clear demarcation, to understand systems and act through models of them. The general design pattern for model based inquiry subsumes the *design patterns* for each of the aspects and calls attention to the coordination among them. More than any of the individual aspects, model-based inquiry highlights the importance of metacognition in moving effectively through cycles of inquiry.

This section draws on the model-based inquiry framework in White and Frederiksen (1998) and White, Shimoda, and Frederiksen (1999). More recently these ideas have been used in simulation environments to support students to carry out investigations, work through inquiry cycles, and build and test models (Clarke-Midura, Code, Zap, & Dede, 2012; Shute et al., 2010; Quellmalz et al., 2012). Providing students with considerable flexibility to choose what to do, when and where, in a simulated microworld, be it in a laboratory, out in the field, under the sea, or on an alien planet, makes it possible to assess their information management and interactive, iterative, reasoning. Capturing log files of actions as rich Work Products makes it possible to evaluate many Observables automatically. This design pattern provides support to designers wishing to assess this overarching aspect of model-based reasoning.

R.J. Mislevy et al., *Assessing Model-Based Reasoning using Evidence-Centered Design*, SpringerBriefs in Statistics, DOI 10.1007/978-3-319-52246-3_11

11.1 Rationale, Focal KSAs, and Characteristic Task Features

The philosophy of science, Giere (1994) argues, assumes that the language of science has a syntax, a semantics, and, finally, a pragmatics. He continues,

> While syntax is deemed important, semantics, which includes the basic notions of reference and truth, has received the most attention. Much of the debate regarding scientific realism, for example, has been conducted in terms of the reference of theoretical terms and the truth of theoretical hypotheses. Pragmatics has been largely a catchall for whatever is left over, but seldom systematically investigated. I now think that this way of conceiving representation in science has things upside down (p. 742).

Model-based reasoning is all about pragmatics. A philosophy of science is not sufficient for either understanding how scientists use models in practice or for how to help students learn to use them; a cognitive psychology of science is required as well. While the preceding sections on aspects of model-based reasoning illuminate important cognitive activities in model-based scientific inquiry, it is the heuristics, the strategies, the procedures, and the self-regulating tools that people need to use models effectively in real-world situations. It is this higher-level, coordinating, or executive level of cognition that the Model-Based Inquiry design pattern addresses.

The Focal KSAs in this design pattern are students' capabilities to manage their reasoning in inquiry cycles. The specific aspects of model-based reasoning discussed in the preceding sections are brought to bear, but is their use coordinated, efficient, coherent, and effective—or is movement through the investigation disjointed, unsystematic, inefficient, and aimless? Are students bringing to bear self-monitoring skills to understand whether model evaluation is needed, or whether a provisional model need to be revised or elaborated?

Any task developed for an overall assessment of model-based reasoning must contain more than one characteristic feature-set from the more specific *design patterns*. As with all of these *design patterns*, there must be a real-world problem being addressed. This problem must require the use of models and/or a modification of models in order to develop an explanation or prediction of some phenomena. The Model-Based Inquiry design pattern goes beyond the specific *design patterns* by addressing information and reasoning across the aspects.

Many of the examples mentioned in the previous sections can be expanded to include multiple aspects of model-based reasoning, and would therefore be instances from the overall design pattern. Stewart and Hafner's genetics curriculum can be thought of as one large assessment task, or it can be broken down into several distinct assessments. In this case, the assessment would start out where the students are applying the simple dominance model to a given situation (as seen in model use). The students then are presented a situation where it does not fit—say, three possible traits instead of two. The students must identify the inadequacies of the simple dominance model (model evaluation) and modify their model (model elaboration.) Students are given further information to lead to more complicated models. At points, they must revise or further elaborate their model in light of new data. Work Products for this overarching task would include the explanations for the models and how they fit the situations, the overall outcomes of using the model to explain or predict behavior, and representations of the models. These Work Products can then be used to evaluate a student's model-based reasoning in the context of modes of inheritance.

Box Robotics-6. Model-Based Inquiry While the preceding discussions of the robotics task have focused on particular aspects of model-based reasoning, it will be clear by now that cycles of design, construction, testing, evaluating, and revising the rover are at the heart of the task. In each phase, reasoning through the underlying gear model and circuit model are required. But the task is structured so as to help the students become aware of the reasoning aspects and the rhythms of such investigations.

The Focal KSA is managing one's work through such cycles, here in the context of generously scaffolded disciplinary content through the MOOC. Additional KSAs are the disciplinary models, the specifics of the circuits, motors, gearboxes, and wheels through which the rovers are constructed, and the proficiency with the necessary tools, representations, and manipulations in a given phase of the investigation. In the simulation phase, these are the tools, affordances, and representations of the simulation environment. In the physical phase, they are proficiencies for the manual planning, assembly of, and operation of the components (plus proficiency of using the laser cutter, if a student is making custom wheels).

We have defined Model-Based Inquiry as an organizing framework for organizing the more specific aspects of model-based reasoning: awareness of those aspects, knowing how they are related, and how to move from one another effectively. The Characteristic Feature for a situation to provide evidence about these capabilities is that it must require two or more aspects of reasoning, and a student must move among them.

An important Variable Feature is the nature and amount of scaffolding that is provided for moving among aspects. The simulation phase in the robotics task provides a good deal of support, in two ways. First, the MOOC materials walk the student through the required background information on the models and the simulation tools and affordances, then structure the initial work in building the first simulation model (Model Formation) and running it (Model Use). Second, the Learning Companion (Fig. R5) provides more specific advice for examining the results of a hill-climbing attempt (Model Evaluation) and offers suggestions on what to try next (Model Revision). As seen in the flowchart, after three unsuccessful tries, it suggests getting help from the outside—an instructor or a friend perhaps—because the inquiry cycles are not converging within the amount of scaffolding the Learning Companion can offer. Note that providing its advice, the Learning Companion is carrying out assessment itself, using the log file Work Product, and counting attempts and comparing attempt results and students' revisions in response to them.

The physical phase offers much less explicit support. The rationale is that after successful completion of the analogous task in the simulation world, a student will have acquired some understanding of the build-run-evaluate-revise inquiry cycle. With less scaffolding, this may or may not happen. Potential Work Products that can provide evidence could include a video capture of the work, an after-the-fact explanation of the work, and a student's running record of models, results, interpretations, and revisions. Note that asking for students to keep a running record with these categories is itself is a mild form of scaffolding. Potential Observations of such Work Products could include the following:

- The degree to which a student organized their activity around such organized cycles.
- Instances of skipping necessary aspects of reasoning, or missing cues as to what actions should be taken next.
- "Churning" activity, with lots of building and running models but no real systematic learning from results and acting to improve on them.

11.2 Additional KSAs

As with the other *design patterns*, the Additional KSAs in the design pattern for assessing model-based inquiry include knowledge of the models, context, and scientific content involved. The mix of these Additional KSAs, if any, that is jointly a target of inference with inquiry itself must be determined in light of the purpose of the assessment and test population. Additional KSAs that are not part of the target of the assessment should be avoided or supported, or the assessor should ascertain that the students are sufficiently familiar with them so that they are not significant sources of difficulty.

11.3 Variable Task Features

Because inquiry tasks encompass the aspects of model-based reasoning addressed so far, all of the Variable Task Features for relevant aspects are open for consideration. This includes the identification and complexity of the model and which tools and representational forms are used. Some design choices can cut across aspects of the larger task (such as the models and content area that are involved) while others (such as scaffolding) can differ from one aspect to another (e.g., a checklist just for model evaluation). Time frame is an important Variable Feature for investigations. Non-trivial investigations can easily take an hour or more, and learning tasks can extend to days or weeks.

Choices regarding the content area will be shaped by the intended purpose of the task. In the classroom or as part of a curriculum, the content is likely based on the models that are the focus of instruction, so the task can pose high demands for this knowledge. The students in the Baxter et al. Mystery Boxes study had just completed a unit in electrical circuits. In a high-stakes accountability test where both the models and the inquiry processes are addressed in the standards, demands for both may be imposed and the Additional KSAs regarding the model and scientific content can be construct-relevant. In a large-scale task that is meant to focus on the inquiry process and not be confounded with content, the models and content can be chosen to be familiar enough to students to minimize poor performance for these reasons. For example, models from middle school standards could be used in a secondary-level task in order to focus its evidentiary value on inquiry.

An important Variable Task Feature in designing inquiry tasks is the degree of scaffolding to provide students as they move from one aspect of an inquiry to another, for managing information, evaluating progress, and deciding what to do next. This self-monitoring is central to inquiry and one of the hardest aspects for students to learn (and for educators to assess). Research on scaffolding students' learning about inquiry holds insights for task designers. In inquiry assessment, more scaffolding is appropriate for earlier learners; it helps them engage meaningfully with the task and ensure that evidence will be obtained for aspects of the investigation. On the other hand, scaffolding the processes means less evidence is available about students' capability to manage their activity in the investigation.

White and Frederiksen (1998) describe a sequence of seven instructional tasks that constitute a middle-school course on mechanics, implemented in the ThinkerTools software. Scaffolding was progressively decreased as students became familiar with inquiry processes and expectations. Associated with each task context is a task document in which students carry out their work. They include a Project Journal, a Project Report, a Project Evaluation, and a System Modification Journal for recording their system modifications and the reasons for them. The documents are organized around a sequence of subtasks (or subgoals) for that task. For example, the Project Journal is organized around the inquiry cycle. The White et al. (1999) simulation environment SCI-WISE additionally provides interactive support in the form of personified "agents":

> In addition to Task Documents, each Task Context has a set of advisors associated with it, including a Head Advisor and a set of Task Specialists. There is a Head Advisor for each Task Context; namely, the Inquirer for doing research projects, the Presenter for creating presentations, the Assessor for evaluating projects, and the Modifier for making changes to the SCI-WISE system. The Head Advisor gives advice regarding how to manage its associated task, suggests possible goal structures for that task, and puts together an appropriate team of advisors. For example, our version of the Inquirer follows the Inquiry Cycle shown in [Fig. 2.2 of this paper]. It suggests pursuing a sequence of subgoals, and each such subgoal has a Task Specialist associated with it, namely, a Questioner, Hypothesizer, Investigator, Analyzer, Modeler, and Evaluator (p. 164).

In computer-based tasks, a developer could choose which agents to make available to examinees and what degree of support they could provide, in order to tailor scaffolding within and between aspects of model-based reasoning during an inquiry task. As always, however, providing tools that support inquiry-related KSAs introduces at the same time a demand for the Additional KSAs to use them effectively.

11.4   Potential Work Products and Potential Observations

Model-based inquiry tasks can be designed to produce Work Products that provide evidence about specific aspects of model-based reasoning within the investigation and/or evidence about managing reasoning across aspects over the course of the investigation. Since aspect-specific Work Products and Potential Observations were discussed previously, after a brief comment, this section focuses on Work Products and Potential Observations that address the encompassing inquiry process.

As mentioned above, all of the potential Work Products that contain evidence about aspects of model-based reasoning can be considered in a fuller inquiry task, and all of the Potential Observations that could be evaluated for these aspects can be considered. In a more detailed scoring scheme, the Observable Variables from the specific aspects can be evaluated and reported separately. This is useful for providing feedback to students in instructional settings: What did they do well in this task, where did they have trouble, and what experiences will help them improve?

Work Products that directly evidence the larger inquiry process must provide information beyond specific aspects of model-based reasoning. This means evidence about the way a student moves through the investigation. One class of Work Products provides some form of trace of the steps a student has taken, such as a video recording, a think-aloud protocol, or a log of actions captured in a computer–based task. The National Board of Medical Examiners' Primum® computer-based diagnostic tests, which are now required for licensure in the United States, capture each step in a solution in a "transaction list." Automated scoring algorithms (more about this below) extract information from the transaction list about both the final solution and selected aspects of the process. In general, less comprehensive Work Products include notebooks, explicit reports of inquiry phases, and written or oral explanations along the way of why certain actions were taken. Oral explanations can be prompted or unprompted. We will say more below about responses to "metacognitive" questions.

Final and intermediate products in an inquiry task are Work Products that can provide indirect evidence about inquiry procedures. A correct solution presumably is more likely to have occurred from effective model-based reasoning, although the efficiency of that reasoning is not available to evaluate from this Work Product alone. The qualities of a final solution to a problem, such as a model proposed for a situation after multiple iterations through the inquiry cycle, can be of interest in and of themselves. Only qualities of the final product may be addressed when the

Table 11.1 Quality of cognitive activity in mystery box solutions (Baxter, Elder, & Glaser, 1996)

Cognitive activity	Range of variation	
	Low	High
Explanation	Single statement of fact or descriptions of superficial features	Principled, coherent
Plan	Single hypothesis	Procedures and outcomes
Strategy	Trial and error	Efficient, informative, goal-oriented
Monitoring	Minimal and sporadic	Frequent and flexible

purpose of an assessment is licensure, for example. But when the purpose is learning, the evaluation of successive provisional models offers clues about the efficiency and appropriateness of successive cycles of model evaluation and revision.

The choice of Work Products to capture is linked to the choice of scaffolding to provide. The task documents White et al. (1999) provided students to record, evaluate, and explain their progress through an investigation not only serve as Work Products, but they support metacognition to manage their activity through the investigation.

What Observable Variables that hold evidence about model-based inquiry can be evaluated from Work Products? Baxter et al. used the Mystery Boxes tasks to study "expertise" in middle school students' inquiry capabilities in a domain known to be familiar to them. Table 11.1 summarizes dimensions of variation they found in a think-aloud protocol and solution-trace Work Products. They are the basis of generic Observable Variables that can be applied more generally in inquiry assessments, as tailored to the processes in the specific investigation.

Baxter et al. evaluated students' investigation procedures by painstakingly parsing "thick" Work Products such as explanations, solution paths, and conversations of thirty-one students. In more complex investigations at larger scales, the amount of rater time and expertise required to carry out such evaluations for these Observable Variables renders them impractical.

An alternative that is available when the investigations are carried out in a computer-based form is automated scoring of solution traces (Bejar, Mislevy, Rupp, & Zhang 2016). In Primum® tasks, low-level features of solutions are identified, combined into higher-level features through logical rules (such as whether efforts to stabilize an emergency patient were carried out first rather than later in the investigation), and evaluated using a regression function that compares them to the high-level features of experts' solutions (Margolis & Clauser, 2006).

More generally, Gobert, Sao Pedro, Baker, Toto, and Montalvo (2012) provide both an overview of approaches to automated scoring of performances on inquiry tasks in simulation environments and examples from their work with *Science Assistments*. The first category they discuss is knowledge engineering/cognitive task analysis approaches, in which rules are defined a priori to encapsulate specific behaviors or differing levels of systematic experimentation skill. The second category is educational data mining/machine learning approaches, in which student inquiry behaviors are discovered from data. Their own examples blend ideas from the two. Leveraging Gobert's previous research on inquiry (including the model-based reasoning research cited above), they designed a simulated laboratory and affordances that minimized construct-irrelevant demands and maximized the evidentiary value of students' actions for how they were managing the inquiry process. For example, they provided a tool using drop-down menus for students to build hypotheses they would then test. The general structure was

> When the [*independent variable*] is [*increased/decreased*], the [*dependent variable*] [*increases/decreases/doesn't change*].

The Work Product produced by filling out the hypothesis is a filled in hypothesis statement–captured in a manner that the system knows exactly what the student has specified. Then, the trace of students' more open-ended actions through the environment of setting up tests, monitoring (or not monitoring) results, and setting up subsequent tests based on previous results (or seemingly not) could be detected by patterns discovered in data mining, based on a subset of actions tagged by expert reviewers. Further, an explanation tool similar to the hypothesis tools was used to capture students' interpretations of what they had done:

> When I changed the [*independent variable*] so that it [*increased/decreased*], the [*dependent variable*] *increased/decreased/didn't change*]. I am basing this on: Data from trial [*trial number from table*] compared to data from trial: [*trial number from table*] this statement [*does support/does not support/is not related to*] my hypothesis.

Together, these Work Products and consequent Observable Variables captured consistencies and inconsistencies, efficiencies, and appropriate stepping through inquiry actions, even though the investigation phases could be accomplished in any numbers of ways.

A class of paired Potential Work Products and Potential Observables that is particularly well-suited to instructional tasks is based on responses to metacognitive questions. These are the questions that students should be learning to ask themselves as they develop their inquiry capabilities. For earlier learners, the answers to these questions provide evidence about the degree to which they are thinking about the appropriate features of their work as it proceeds. Their very presence helps the students learn that these are questions that are important in inquiry, and they come to internalize them as they gain experience. For example, White and Frederiksen (1998) acquaint students with a concept they called "Being Systematic": "Students are careful, organized, and logical in planning and carrying out their work. When problems come up, they are thoughtful in examining their progress and deciding whether to alter their approach or strategy." As a Work Product, students rate their own solutions with respect to how systematic they were, on a 1-to-5 scale from "not adequate" to "exceptional."

11.5 Some Connections with Other Design Patterns

Model-based inquiry is an encompassing activity that draws repeatedly and cyclically on more specific aspects of model-based reasoning. When designing an inquiry task, a test developer can use this design pattern to consider the characteristics of Task Features and Work Products that will provide evidence about the movement in the larger space, and the specific *design patterns* to ensure that evidence is elicited as needed about details of the investigation.

The iterative testing and repairing that characterizes troubleshooting can be viewed as a special case of model-based inquiry. Steinberg and Gitomer's (1996) troubleshooting tasks in the hydraulic system of the F-15 aircraft, for example, required iterative cycles of model use, model evaluation, and model revision, with the efficiency of diagnostic tests at the crux of evaluation. The efficiency of tests for evaluating a model becomes particularly important in these more complex tasks. Efficiency is intimately related to understanding both the system in question and the tests that can be carried out, both Additional KSAs that are required jointly for effective troubleshooting. Frezzo, Behrens, and Mislevy (2009) showed how *design patterns* for creating troubleshooting tasks in network engineering are used in the Cisco Networking academy. Seibert, Hamel, Haynie, Mislevy, and Bao (2006) presented a more general design pattern that encompasses troubleshooting, called "Hypothetico-Deductive Problem Solving in a Finite Space."

Chapter 12
Conclusion

Abstract Research on science learning increases our understanding of the capabilities we want to help students develop, and advances in technology expand the ways we can support and assess their learning. Familiar testing practices offer little guidance, however, for designing valid assessments of more ambitious proficiencies in more complex settings. These *design patterns* can support the development of tasks for assessing model-based reasoning in a variety of contexts, including standards-based assessment, classroom assessment, large-scale accountability testing, and simulation- and game-based assessment.

Model-based reasoning, and inquiry in general, are both increasingly important and difficult to assess (Means & Haertel, 2002). Assessing factual knowledge and isolated procedures is easier and more familiar—and not surprisingly, constitutes the bulk of current science assessment. The *design patterns* developed here can be used as starting points for building assessment tasks that engage more deeply with model-based reasoning. Task developers can determine which aspects of model-based reasoning to address and use the corresponding *design patterns* to make them aware of design choices and support their thinking about how to make them. The *design patterns* are organized around elements of an assessment argument structure as it has emerged from research on assessment design and validity theory. In this way, the *design patterns* leverage both research on model-based reasoning and practical experience in assessment design in this area, in a form that is specifically designed to support task developers.

12.1 Standards-Based Assessment

As part of the standards-based reform movement over the last two decades, states and national organizations have developed content standards outlining what students should know and be able to do in core subjects, including science (e.g., NRC, 1996, 2012). These efforts are an important step toward furthering professional consensus about the knowledge and skills that are important for students to learn at

© The Author(s) 2017
R.J. Mislevy et al., *Assessing Model-Based Reasoning using Evidence-Centered Design*, SpringerBriefs in Statistics, DOI 10.1007/978-3-319-52246-3_12

various stages of their education. They are the basis of states' large-scale accountability tests, as was the case under the requirements of the 2001 No Child Left Behind (Public Law 107-110, 2002) legislation and is currently advocated by the Next Generation Science Standards (NGSS Core States, 2013).

But standards in their current form are not specifically geared toward guiding assessment design. A single standard for science inquiry will often encompass a broad domain of knowledge and skill, such as "develop descriptions, explanations, predictions, and models using evidence" (NRC, 1996, p. 145) or "communicate and defend a scientific argument" (p. 176). They stop short of laying out the inter-connected elements that one must think through to develop a coherent assessment: the competencies that one is interested in assessing, what one would want to see students do as evidence of those competencies, and assessment situations that would elicit such evidence. Even NGSS performance expectations (NGSS Core States, 2013), which sketch illustrative tasks that could elicit processes, disciplinary knowledge, and overarching concepts, provide little guidance for task developers to operationalize the ideas at scale, with reliability and validity.[1]

Design patterns bridge knowledge about aspects of science inquiry that one would want to assess and the structures of a coherent assessment argument, in a format that guides task creation and assessment implementation. The focus at the *design pattern* level is on the substance of the assessment argument rather than on the technical details of operational elements and delivery systems. Thinking through the substance of assessment arguments for capabilities such as model-based reasoning and inquiry promotes the goals of efficiency and validity. It enables test developers to go beyond thinking about individual assessment tasks and to instead see instances of prototypical ways of getting evidence about the acquisition of various aspects of students' capabilities.

Design patterns bring insights from cognitive psychology, science education, and the philosophy of science together in a form that can support designing assessment tasks for both classroom and large-scale assessments. It is a particular advantage of *design patterns* to center on aspects of scientific capabilities, as opposed to task formats or assessment purposes. The essence of the capabilities and building assessment arguments around them is seen as common, with options for tailoring the details of stimulus situations and Work Products to suit the particulars of a given assessment application.

12.2 Classroom Assessment

Design patterns built around national or state science standards constitute a sta-tionary point to connect both classroom and large-scale assessment with develop-ments in science education and educational psychology. There is often a disjuncture

---

[1]In the terminology of *design patterns*, NGSS Performance Expectation highlight instantiations of Characteristic Features and Variable Features that would tap Focal KSAs.

between classroom assessment and large-scale assessment; *design patterns* help make it clear that it is the same capabilities being addressed in both, although the assessments reflect different design choices about such features as time, interactivity, and Work Products to accommodate the different purposes and constraints of large-scale and instructional tests.

Truly "knowing" models in science is more than echoing concepts and applying procedures in isolation; it is using models to do things in the real world: reasoning about situations through models; selecting, building and critiquing models; working with others and with tools in ways that revolve around the models. Students develop these capabilities by using them, first in supported activities that make explicit the concepts, the processes, and the metacognitive skills for using them. It is no coincidence that most of the examples we have used to illustrate science *assessment* are drawn from projects that focus on science *learning*. These *design patterns* for assessing model-based reasoning can help make the advances in science education more accessible to classroom teachers and curriculum developers as well as to researchers and assessment professionals.

12.3   Large-Scale Accountability Testing

The changing landscape of large-scale accountability assessments places extraordinary demands on state and local education agencies. No Child Left Behind legislation required large-scale testing at the level of the state, with attendant needs for efficient administration, scoring, linking of forms, and cost-effective development of assessment tasks at unprecedented scales. Tasks must address states' content standards. At the same time, educators want tasks that assess higher-level skills and are consistent with both instructional practice and learning science.

It is widely accepted that more complex, multi-part assessment tasks are better suited to measuring higher-level skills. But their cost and incompatibility with conventional test development and implementation practices stand in the way of large-scale use. Many states and their contractors have turned to computer-supported assessment task development and delivery to help them meet these challenges. For large-scale assessments, technology-based tasks such as simulations and investigations to address higher-level skills and support learning have proved difficult and costly to develop, especially when employing procedures that evolved from conventional multiple-choice item development practices.

Traditionally, items for large-scale assessments are developed by item writers who craft each item individually. Often as many as half of the items do not survive review. This low survival rate is tolerable because of the relatively low cost of developing individual multiple-choice items. It is not economical for developing the more complex tasks needed to address higher-level skills. Moreover, the thought and problem solving invested in developing any particular item is tacit in conventional item development procedures. The thinking invested, the design challenges met, and the solutions reached remain undocumented and inaccessible to help item writers develop additional items. This process is untenable in the long run for tasks that require an order of magnitude more time and resources than multiple-choice items.

Design patterns are part of the solution. A *design pattern* specifies a design space of interconnected elements to assemble into an assessment argument. This design space focuses on the science being assessed and guides the design of tasks with different forms and modes for different situations. *Design patterns*, in turn, ground *templates* for authoring more specific families of tasks.

In the context of large scale accountability assessments, *design patterns* thus fill a crucial gap between broad content standards and implemented assessments tasks, in a way that is more generative than test specifications and which addresses alignment through construction rather than retrospective classification. The time and analysis invested in creating *design patterns* eliminates duplicative efforts of re-addressing the same issues task by task, program by program. *Design patterns* can be developed collaboratively and shared across testing programs. Each program can construct tasks which, by virtue of the pattern, address key targets in valid ways, but make design choices that suit their particular constraints and purposes. Thus, *design patterns* add value not just for local development but for accumulating experience and debating standards in the state, national, and international arenas.

12.4   Simulation- and Game-Based Assessment

The ability to create computer-based simulation environments has opened the door to assessing model-based reasoning in complex, interactive environments practically anytime, anywhere. Simulations have a great advantage of making visible and more amenable to the cognitive aspects of modeling phenomena that might be too small, too big, too costly, too distant, or too dangerous. They can provide facsimiles of the tools and representations real scientists and engineers use. They can provide scaffolding, supporting material, just in time information, and provide feedback to students as they work as well as informing teachers or more distant users. And perhaps most importantly, they allow for interactions with situations—a hallmark of model-based reasoning in action.

Furthermore, log files of actions, time-stamps, and plans and products can all be captured automatically as rich, detailed Work Products. An exciting frontier of educational assessment is developing automated methods of evaluating log files, through which cognitively meaningful patterns and features of work are detected and characterized as Observations. There is great potential for many kinds of assessments: From moment-to-moment assessment and feedback in simulations for learning, to simulated investigations embedded in curricula, to technology-enhanced tasks in large-scale assessments.

But it is difficult to design tasks to be valid, comparable, and fair. Design patterns help with some of the thorniest problems—such as identifying alternative explanations for poor performance, and adapting Variable Task Features and Work products to a testing population, and addressing higher-level process skills such as

Model Revision within disciplinary contexts that match students' instructional backgrounds.

Design patterns thus offer support to designers of complex, computer-based tasks, to help make sure that they can produce not only good simulation around valued disciplinary content and processes, but valid assessment evidence as well. Resources on how to use ECD more generally, and *design patterns* in particular, in designing game-based and simulation-based assessments are now beginning to appear (e.g., Clarke-Midura & Dede, 2010; Gobert, Sao Pedro, Baker, Toto, & Montalvo, 2012; Mislevy et al., 2013; Riconscente, Mislevy, & Corrigan, 2015; Shute, Ventura, Bauer, & Zapata-Rivera, 2009).

12.5   Closing Comments

Model-based reasoning is central to science. Research from a sociocognitive perspective on the nature of model-based reasoning and how people become proficient at using it is beginning to revolutionize science education. Assessment is integral to learning, not just for guiding learning but for communicating to students and educators alike just what capabilities are important to develop, and how to know them when we see them. But the interactive, complex, and often technology-based tasks that are needed to assess model-based reasoning in its fullest forms are difficult to develop. The suite of *design patterns* to support the creation of tasks to assess model-based reasoning hold promise to help bring assessment into line with contemporary views of science learning and science assessment.

Appendix
Summary Form of Design Patterns for Model-Based Reasoning

R.J. Mislevy et al., *Assessing Model-Based Reasoning using Evidence-Centered Design*, SpringerBriefs in Statistics, DOI 10.1007/978-3-319-52246-3

	Model formation	Model use	Model elaboration	Model articulation
Summary	This design pattern supports developing tasks in which students create a model of some real-world phenomenon or abstracted structure, in terms of entities, structures, relationships, processes, and behaviors.	This design pattern supports developing tasks that require students to reason through the structures, relationships, and processes of a given model.	This design pattern supports developing tasks in which students elaborate given scientific models by combining, extending, adding detail to a model, and/or establishing correspondences across overlapping models.	Tasks supported by this design pattern assess students' ability to articulate the meaning of physical or abstract systems across multiple representations. Representations may take qualitative or quantitative forms. This DP is relevant in models with quantitative and symbolic components (e.g., connections between conceptual and mathematical aspects of physics models).
Rationale	Ability to build a model is a fundamental component of inquiry-based science. During the construction of a model, students make design decisions regarding the question(s) they are interested in answering, what variables they need to include, how "precise" their model needs to be, and how it corresponds to	Scientific models are abstracted schemas involving entities and relationships, meant to be useful across a range of particular circumstances. Procedures within the model space can be carried out to support inferences about the situation beyond what is immediately observable.	Like scientists, students of science should be familiar with the processes that lead to the development of scientific theories and the situated use of scientific models. Model elaboration, in which existing models are combined or extended to incorporate new data or to increase theoretical parsimony, is one such	Scientists reason through problems both as qualitative or physical relationships and as symbolic systems. This ability to articulate across multiple qualitative and/or quantitative representations or physical realities is crucial to students' development of scientific knowledge and ability.

(continued)

(continued)

	Model formation	Model use	Model elaboration	Model articulation
	the elements of the situation.		aspect of scientific inquiry: The user extends adapts, and connects models as prompted by the target situation.	
Focal KSAs (Note: "ability" here means capability to reason as described in a given context with given models. No claim is made for "abilities" as decoupled from particular models.)	Ability to pose relevant questions about system to construct a model. Ability to relate elements of a model to features of a situation and vice versa. Ability to describe (i.e., narrate) the situation through the entities and relationships of the model. Ability to identify which aspects of the situation to address and which to omit. Decision-making regarding scope and grain-size of model, as appropriate to the intended use of the model.	Ability to reason through the concepts and relationships of a given model to make explanations, predictions and conjectures: • Qualitative reasoning through the model. • Quantitative reasoning through the symbolic representations associated with the model.	Ability to identify links between similar models (that share objects, processes, or states). Ability to link models at different levels or focusing on different aspects of phenomena to create a larger, more encompassing model.	Ability to articulate meanings between qualitative and/or quantitative systems associated with scientific phenomenon. Ability to transform information between qualitative and/or quantitative systems associated with scientific phenomenon.
Characteristic features	Specific situation or data (either provided or previously generated by student), to be modeled, for some purpose. Correspondence must be established between	Real-world situation and one or more models appropriate to the situation. Focus is on reasoning through the schema and relationships embedded in the model. Reasoning is as	Real-world situation and one or more models appropriate to the situation, for which details of correspondence need to be fleshed out. Addresses correspondence between	Multiple inter-related representation systems. Task addresses relationship in expressions from one system to another.

(continued)

(continued)

	Model formation	Model use	Model elaboration	Model articulation
	elements of the model and elements of the situation.	if the model is appropriate to the situation is the focus.	situation and models, and models with one another. Problem solution involves combining or making additions to existing models by, for example, embedding a model in a larger system, adding more parts to the model, or incorporating additional information about a real-world situation into the schema the model represents.	
Add'l KSAs	Familiarity with real-world situation. Knowledge of model at issue. Domain area knowledge (declarative, conceptual, and procedural). Familiarity with required modeling tool(s) (e.g., STELLA, Marshall's arithmetic schema interface). Familiarity with required symbolic representations associated procedures (e.g.,	Familiarity with real-world situation. Knowledge of model at issue. Domain area knowledge (declarative, conceptual, and procedural). Familiarity with required modeling tool(s). Familiarity with required symbolic representations, associated procedures. Familiarity with task type (e.g., materials, protocols, expectations).	Familiarity with real-world situation. Knowledge of model at issue. Domain area knowledge (declarative, conceptual, and procedural). Familiarity with required modeling tool(s). Familiarity with required symbolic representations, associated procedures. Familiarity with task type (e.g., materials, protocols, expectations).	Knowledge of and ability to reason within qualitative and quantitative systems implied in the task. That is, this DP isolates the ability to move *between* systems, and therefore it presupposes students' ability to operate *within* the symbolic etc. systems involved. Knowledge of model at issue.

(continued)

(continued)

	Model formation	Model use	Model elaboration	Model articulation
	Marshall's schema forms, mathematical notation). Familiarity with task type (e.g., materials, protocols, expectations).			Domain area knowledge (declarative, conceptual, and procedural). Familiarity with required modeling tool(s). Familiarity with required symbolic representations, associated procedures. Familiarity with task type (e.g., materials, protocols, expectations).
Variable features	Is problem context familiar (i.e., degree of transfer required)? To what degree is the model prompted? Is model formation isolated, or in the context of a larger investigation? Complexity of problem situation; e.g., simplified situation versus more factors and realism; simple mapping of word problem quantities, as in Marshall's schemas for arithmetic, versus problem requiring large, hierarchical, and/or sequential modeling.	Is problem context familiar (i.e., degree of transfer required)? Is model use isolated, or in the context of a larger investigation? Complexity of model. Complexity of situation. Complexity of reasoning required (e.g., number of variables in model, number of steps required)? Model provided or generated by student? Data provided or generated by student? Degree of scaffolding provided (especially if	Is problem context familiar (i.e., degree of transfer required)? Model given to student(s), versus model to elaborate produced by student(s) themselves. Complexity of elaboration required; e.g., minor modification of familiar model, versus nesting of models, versus elaboration to new previously unknown model. Is model use isolated, or in the context of a larger investigation? For example, must	Articulation between semantic and symbolic systems, among different symbolic systems, or across multiple systems? Is problem context familiar (i.e., degree of transfer required)? Number of systems used (model, symbolic, physical, abstract). Complexity of systems. Complexity of mappings (conditions, # issues to simultaneously consider). Prior exposure to representations and mapping conventions.

(continued)

(continued)

	Model formation	Model use	Model elaboration	Model articulation
	Complexity of the model; i.e., number of variables, complexity of variable relations, number of representations required, whether the model is runnable). Well-defined problem versus ill-defined problem (or gradations thereof). Is extraneous information provided (makes tasks more difficult)? Kind of model needed for problem goal: simple and quick versus more exact and complex. Role/depth of approximation required. Degree of scaffolding provided. Group or individual work?	model use involves strategies, alternate approaches, and multiple steps). Group or individual work?	experimental work or supporting research be carried out in order to ground the elaboration? Single model to elaborate versus establishing a correspondence among models at different levels or with different foci? Degree of scaffolding provided (e.g., is need for elaboration prompted? Are hints or checklist provided to guide elaboration?) Group or individual work?	Is articulation the focus of a task, or is it part of a larger task? If part of a larger task, is the articulation problem presented to the student or is the need for, and carrying out of, articulation to be sought in the trace of a free-form solution trace? Degree of scaffolding provided (e.g., is need for elaboration prompted? Are hints or checklist provided to guide elaboration?) Group or individual work?
Potential work products	Final version of model. Explanation of model interpretation (as written or typed log or response, or audio transcript). Correspondence mapping between elements or	Selection of hypotheses, predictions, retrodictions, explanations, and/or missing elements of real world situation. Constructed hypotheses, predictions, retrodictions,	Correspondence mapping between elements or relationships of model and real-world situation. Correspondence mapping between elements or	Re-expression of information in one or more systems in terms of another system. Cross-system problem solutions with mappings (e.g., force diagrams and

(continued)

(continued)

	Model formation	Model use	Model elaboration	Model articulation
	relationships of model and real-world situation when there are gaps in one or the other to be filled. Notes taken during model building process. Trace of steps taken to build model. See Scalise and Gifford (2006) for a taxonomy and illustrations of work product formats that are amenable to automated scoring.	explanations, and/or missing elements of real world situation, via • Creation of one or more representational forms. • Filling in given, possibly partially filled in, representational forms. Intermediate products developed in selection/construction of hypotheses, predictions, explanations, and/or missing elements. Written/oral explanation of the hypotheses, predictions, explanations, and/or missing elements. Trace of actions taken in solution. Talk-aloud of solution. Critique of a given solution.	relationships of overlapping models. Final elaborated model (Physical, symbolic, verbal, etc., as appropriate). Trace of steps and provisional models. Written/oral explanation of reasoning behind elaboration.	equations). Can be prompted with "show your work." Verbal descriptions and explanations of meanings across representational systems. Predictions for one system given information about an associated system. Selection of system for scenario presented in terms of other systems.
Potential observations	Qualities of final model: • Accuracy of determination of what variables are important to include in the model.	Quality of students' explanations, predictions, or retrodictions as reasoned through the model; e.g., correctness, appropriateness	Extent to which student catenates models appropriately across levels; that is, common entities and processes match up appropriately (e.g.,	Quality of operations applied *across* systems. Extent to which student accurately maps one system into another, rather than back onto itself.

(continued)

(continued)

Model formation	Model use	Model elaboration	Model articulation
• Accuracy of representation of relations among variables. • If runnable model, quality of model output. • Appropriateness of degree of precision of model of phenomenon or system being modeled. Modeling process: • Efficiency in terms of tools and representations. • Quality (relevancy, accuracy) of questions asked about the system to inform the construction of a model. Includes domain-specific heuristics and domain-specific explanatory schemas when these are targets of inference. • If talk-aloud, degree to which student talks about the meaning of the data. • Quality of rationale student provides for steps in construction of model.	(i.e., quality of the product of model use) Qualities of solution procedure, such as appropriateness, efficiency, systematicity, quality of strategy, and effectiveness of procedures (i.e., qualities of the student's process). Quality of student's explanation of her own solution through a model (i.e., quality of the student's explanation of their process of model use, as distinct from the quality of the product of their reasoning).	individual-level and species-level models in transmission genetics). Quality of student explanation of modifications, in terms of features of data/purpose that require reasoning across levels/submodels. Accuracy and completeness of mappings between a real-world situation and elaborated model.	Accuracy of predictions in system y based on expressions in system x. Accuracy and completeness of creation of system y based on expressions in system x. Quality/appropriateness of description of meaning of information across systems. Accuracy of selection of system (given example—i.e., instruction would have made the various systems explicit to students).

(continued)

(continued)

	Model formation	Model use	Model elaboration	Model articulation
	Includes domain-specific heuristics and domain-specific explanatory schemas when these are targets of inference. • Quality of rationale for what entities and relationships are expressed in the model, versus those which are omitted. • Speed at which student forms model [indicates automaticity; Kalyuga 2006].			
Selected references	diSessa (1983) Hunt and Minstrell (1994) Kintsch and Greeno (1985) Kindfield (1999) Redish (2004)	Stewart and Hafner (1994) Johnson-Laird (1983) Gentner and Stevens (1983) Hestenes, Wells, and Swackhamer (1992)	Marshall (1993, 1995) Stewart and Hafner (1994) Frederiksen and White (1988)	Greeno (1989) diSessa (1983, 1993) Marshall (1993)

	Model evaluation	Model revision	Model-based inquiry
Summary	This design pattern supports developing tasks in which students evaluate the correspondence between a model and its real-world counterparts, with emphasis on anomalies and important features not accounted for in the model.	This design pattern supports developing tasks in which students revise a model in situations where a given model does not adequately fit the situation or is not sufficient to solve the problems at hand.	This design pattern supports developing tasks in which students work interactively between physical realities and models, using principles, knowledge and strategies that span all aspects and variations of model-based reasoning.
Rationale	It is essential to be able to determine the degree to which a model is in fact an appropriate method for reasoning about a physical situation. Examining evidence of the nature and quality of model misfit to data is important in this regard. Key to this endeavor is the degree to which predications and inferences from the model, given some of the data, are consistent with other parts of the data or further data that could be gathered.	Model-based reasoning concerns making inferences about real-world situations through the entities and structures of a model. When the model is not appropriate for the job at hand, either because it does not fit or it does not adequately capture the salient aspects of the situation, it is necessary to be able to revise the model.	Coordinating the aspects of model-based reasoning in inquiry requires not only being able to reason in each of the specific aspects, but to coordinate their use in iteratively building and testing models as inquiry proceeds. Metacognitive aspects of strategies for model-based reasoning are involved.
Focal KSAs (Note "ability" here means capability to reason as described in a given context with given models. No claim is made for "abilities" as decoupled from particular models.)	In broad terms, ability to determine the appropriateness of a model for reasoning about a situation, for a given purpose. More specifically, ability to identify salient features of available data for comparison, and detect anomalies that available models cannot explain.	Ability, in a given situation, to modify a given model so that its features better match the features of that situation for the purpose at hand. More specifically: • Recognizing the need to revise a provisional model. • Modifying it appropriately and efficiently.	Ability to carry out aspects of model-based reasoning when appropriate in an investigation, moving from one to another and using the results of each step to guide the next. Ability to monitor progress and results in inquiry cycle investigations (metacognition, self-regulation).

(continued)

(continued)

	Model evaluation	Model revision	Model-based inquiry
		• Justifying the revisions in terms of the inadequacies of the provisional model.	Ability to take appropriate action with regard to model-based inferences in light of real-world feedback.
Characteristic features	A model is proposed for a situation, and its suitability must be evaluated —is it satisfactory for the purpose? Where and how might it not appear to be adequate?	A situation to be modeled, a provisional model that is inadequate in some way, and the opportunity to revise the model in a way that improves the fit.	Necessity to probe a problem by invoking or revising models to explain phenomena or make predictions. One or more inquiry cycles, and more than one aspect of inquiry, is required, so that managing interaction among them is required.
Add'l KSAs	Familiarity with real-world situation. Domain area knowledge (declarative, conceptual, and procedural). Familiarity with required modeling tool(s). Familiarity with required symbolic representations associated procedures (especially statistical methods). Familiarity with task type (e.g., materials, protocols, expectations). Familiarity with standards of quality and expectation in the field. Ability to encode and represent evidence to be evaluated as an entity	Ability to detect anomalies not explained by existing model (i.e., model evaluation). Familiarity with real-world situation. Domain area knowledge (declarative, conceptual, and procedural). Familiarity with required modeling tool(s). Familiarity with required symbolic representations associated procedures. Familiarity with task type (e.g., materials, protocols, expectations). Ability to engage in model use	Familiarity with real-world situation. Domain area knowledge (declarative, conceptual, and procedural). Familiarity with required modeling tool(s). Familiarity with required symbolic representations associated procedures. Familiarity with task type (e.g., materials, protocols, expectations). Familiarity with standards of quality and expectation in the field. Additional KSAs as might be required in particular aspects of model-based reasoning addressed in design patterns for those aspects.

(continued)

(continued)

	Model evaluation	Model revision	Model-based inquiry
	distinct from representations of the model.		
Variable features	Is the model-to-be-evaluated given, or was it developed by the student in the course of an investigation? Does the situation itself provide feedback about a model (e.g., as in interactive tasks such as troubleshooting)? Is the model satisfactory or not satisfactory? If the model is not satisfactory, in what way(s) is this so? (e.g., lack of fit to observations, inappropriateness to project goal, wrong grain size or aspects of phenomenon). Is problem context familiar (i.e., degree of transfer required)? To what degree is the model evaluation prompted? Complexity of problem situation. Complexity of the model; i.e., number of variables, complexity of variable relations, number of representations required, whether the model is runnable). Is extraneous information present (makes tasks more difficult, because it evokes the need to evaluate	Is the model-to-be-revised given, or was it developed by the student in the course of an investigation? In what way is the model unsatisfactory: Lack of fit to observations, inappropriateness to project goal, wrong grain size or aspects of phenomenon? Are the unsatisfactory aspects provided to the student, or to be discovered through model evaluation? Is model revision iterative, with feedback? To what degree is the model revision prompted? Is problem context familiar? Complexity of problem situation. Complexity of the model; i.e., number of variables, complexity of variable relations, number of representations required, whether the model is runnable). Group or individual work?	Amount of scaffolding provided for working through inquiry cycles (from being walked through cycles, to support within cycles, to unsupported). Note: Scaffolding can be in the form of structured work products. Single cycle or multiple cycles anticipated? Problem given or self-determined in accordance with some criteria? Is problem context familiar? Time scale (e.g., relatively short snippets of fuller investigations as might appear in a large-scale test; short hands-on or simulation tasks, say 30-90 min; investigations that are carried out over days or weeks). Opportunities to engage in persuasion of peers in defense of their own solutions? Group or individual work? Variable features as might be required in particular aspects of model-based reasoning addressed in design patterns for those aspects.

(continued)

(continued)

	Model evaluation	Model revision	Model-based inquiry
	whether certain aspects of the situation should not be modeled)? Group or individual work?		Trace of models as constructed/revised during investigation. Explanations of steps taken during an investigation, moving across inquiry steps. Audio-recordings or transcripts of what students said as they "thought aloud" while solving problems. Computer-kept or notebook records of inquiry steps. Filling out forms summarizing work by step or category. Note: Structured Work Products can be a form of scaffolding. Work products as might be required in particular aspects of model-based reasoning addressed in design patterns for those aspects.
Potential work products	Verbal/written explanation of model-fitting actions [especially looking for prediction of new observations]. Trace of actions. Statements of hypotheses that motivate evaluation procedures. Talk-aloud trace [which may exhibit evidence of model evaluation actions/reasoning]. Representations and summaries of formal model-fitting tools such as statistical tests. Record of results of model-fit. Analyses on forms provided (note: this is a form of scaffolding). Explanation of results of model-fit analysis. Record of hypotheses formulated and tested.	Choice or production of revised model. Explanation of reasoning for revised model. Trace of models as constructed/revised (e.g., sequence of Genetics Construction Kit (GCK) models). Recordings or transcripts of what students said as they "thought aloud" while revising model. Computer-kept records of inquiry steps in which model revision steps are embedded. Notes written by students during model revision.	

(continued)

(continued)

	Model evaluation	Model revision	Model-based inquiry
			Final and intermediate products of solution can hold indirect evidence about inquiry capabilities. Final products that illustrate appropriate models and conclusions suggest appropriate model formation and use at a minimum. Increasingly improved intermediate products suggest appropriate model evaluation and model revision or elaboration.
Potential observations	Comprehensiveness and appropriateness of methods of assessing model fit. Systematicity of model evaluation procedures (e.g., are results of a test used to guide the choice of the next test?). Degree of integration of results from multiple tools/views for assessing model fits. Quality of explanation of model fit. Indication of which aspects of model do not fit, with respect to aspects of data and aspects of model.	Quality and appropriateness of model revisions in order to address inadequacies of provisional model. Degree of and appropriateness of general and/or domain-specific heuristics students use to revise their models. Quality of the basis on which students decide that a revised model is adequate. Quality of explanation of the basis on which students decide that a revised model is adequate.	Presence and quality of activity that reflects self-regulation, including • Explanation of strategy. • Planning. • Reaction to feedback from other people or the situation itself. Quality of student's explanation of her own solution through a model (i.e., quality of the student's explanation of their process of model use, as distinct from the quality of the product of their reasoning).

(continued)

(continued)

	Model evaluation	Model revision	Model-based inquiry
	Quality of determination of whether model misfit will degrade target inferences.	Efficiency of the process by which students evaluate existing models as deficient and revised models as adequate, including use of optimal strategies, sequence, monitoring … This observable can be applied when model revision is part of a larger investigation. Extent to which students extract true results from a set of false models and recognize them as independent of the specific assumptions that vary across the (false) models.	
Selected references	Belsley, Kuh, and Welch (1980) Cartier (2000) Mosteller and Tukey (1977)	Mosteller and Tukey (1977) Rumelhart and Norman (1977) Stewart and Hafner (1991, 1994) Stewart, Hafner, Johnson, and Finkel (1992)	Mosteller and Tukey (1977) Stewart, Hafner, Johnson, and Finkel (1992) Frederiksen and White (1988)

References

Alexander, C., Ishikawa, S., & Silverstein, M. (1977). *A pattern language: Towns, buildings, construction*. New York, NY: Oxford University Press.

Allen, D., Kling, G., & van der Pluijm, B. (2005). Global change curriculum, Unit 1a: Introduction to systems dynamic modeling with STELLA. Ann Arbor, MI: Global change Program, University of Michigan. Retrieved August 2, 2007 from http://www.globalchange.umich.edu/globalchange1/current/labs/Lab2/Intro_Stella.htm

Almond, R. G., Steinberg, L. S., & Mislevy, R. J. (2002). Enhancing the design and delivery of assessment systems: A four-process architecture. *Journal of Technology, Learning, and Assessment, 1*(5). http://www.bc.edu/research/intasc/jtla/journal/v1n5.shtml

Azevedo, R., & Cromley, J. G. (2004). Does training on self-regulated learning facilitate students' learning with hypermedia? *Journal of Educational Psychology, 96,* 523–535.

Baxter, G., Elder, A., & Glaser, R. (1996). Knowledge-based cognition and performance assessment in the science classroom. *Educational Psychologist, 31*(2), 133–140.

Baxter, G., & Mislevy, R. (2005). *The case for an integrated design framework for assessing science inquiry (PADI Technical Report 5)*. Menlo Park, CA: SRI International.

Bejar, I. I., Mislevy, R. J., Rupp, A. A., & Zhang, M. (2016). Automated scoring with validity in mind. In A. Rupp & J. Leighton (Eds.), *Handbook of cognition and assessment*. Hoboken, NJ: Wiley-Blackwell.

Belsley, D. A., Kuh, E., & Welch, R. E. (1980). *Regression diagnostics: Identifying influential data and source of collinearity*. New York: John Wiley.

Bransford, J. D., & Schwartz, D. L. (1999). Rethinking transfer: A simple proposal with multiple implications. *Review of Research in Education, 24,* 61–100.

Buckley, B. C. (2008). Model-based teaching. In N. M. Seel (Ed.), *Encyclopedia of the Sciences of Learning* (pp. 2312–2315). New York: Springer.

Cartier, J. (2000). *Assessment of explanatory models in genetics: Insights into students' conceptions of scientific models (Research Report No. 98-1)*. Madison, WI: National Center for Improving Student Learning and Achievement in Mathematics and Science.

Chi, M. T. H. (2005). Common sense conceptions of emergent processes: Why some misconceptions are robust. *Journal of the Learning Sciences, 14,* 161–199.

Chi, M. T. H., Feltovich, P., & Glaser, R. (1981). Categorization and representation of physics problems by experts and novices. *Cognitive Science, 5,* 121–152.

Clarke-Midura, J., Code, J., Zap, N., & Dede, C. (2012). Assessing science inquiry in the classroom: A case study of the virtual assessment project. In L. Lennex & K. Nettleton (Eds.), *Cases on inquiry through instructional technology in math and science: Systemic approaches* (pp. 138–164). New York, NY: IGI.

Clarke-Midura, J., & Dede, C. (2010). Assessment, technology, and change. *Journal of Research on Technology in Education, 42*(3), 309–328.

© The Author(s) 2017

R.J. Mislevy et al., *Assessing Model-Based Reasoning using Evidence-Centered Design*, SpringerBriefs in Statistics, DOI 10.1007/978-3-319-52246-3

Clement, J. (1989). Learning via model construction and criticism: Protocol evidence on sources of creativity in science. In J. A. Glover, R. R. Ronning, & C. R. Reynolds (Eds.), *Handbook of creativity: Assessment, theory and research* (pp. 341–381). New York: Plenum Press.

Clement, J. (2000). Model based learning as a key research area for science education. *International Journal of Science Education, 22,* 1041–1053.

Collins, A., & Ferguson, W. (1993). Epistemic forms and epistemic games: Structures and strategies to guide inquiry. *Educational Psychologist, 28,* 25–42.

Corcoran, T., Mosher, F. A., Rogat, A. (2009). *Learning progressions in science: An evidence-based approach to reform.* Teachers College-Columbia University: Consortium for Policy Research in Education.

DeBarger, A. H., Krajcik, J. S., Harris, C. J., & Penuel, W. R. (2013). *Designing NGSS assessment to evaluate the efficacy of curriculum interventions.* Princeton, N.J.: Educational Testing Service K-12 Center.

DiCerbo, K., Bertling, M., Stephenson, S., Jia, Y., Mislevy, R. J., Bauer, M., et al. (2015). The role of exploratory data analysis in the development of game-based assessments. In C. S. Loh, Y. Sheng, & D. Ifenthaler (Eds.), *Serious games analytics: Methodologies for performance measurement, assessment, and improvement* (pp. 319–342). New York: Springer.

diSessa, A. A. (1983). Phenomenology and the evolution of intuition. In D. Gentner & A. Stevens (Eds.), *Mental models* (pp. 15–33). Hillsdale, NJ: Lawrence Erlbaum Associates, Inc.

diSessa, A. A. (1993). Towards an epistemology of physics. *Cognition and Instruction, 10,* 105–225.

Duncan, R. G. (2006). The role of domain-specific knowledge in promoting generative reasoning. In *Proceedings of the 7th international conference on learning sciences* (pp. 147–153). Bloomington, Indiana: International Society of the Learning Sciences.

Fauconnier, G., & Turner, M. (2002). *The way we think.* New York: Basic Books.

Frederiksen, J. R., & White, B.Y. (1988). Implicit testing within an intelligent tutoring system. *Machine-Mediated Learning, 2,* 351–372.

Frezzo, D. C., Behrens, J. T., & Mislevy, R. J. (2009). Design patterns for learning and assessment: facilitating the introduction of a complex simulation-based learning environment into a community of instructors. The *Journal of Science Education and Technology.* Retrieved January 27, 2010, from http://www.springerlink.com/content/566p6g4307405346/

Frigg, R. & Hartmann, S. (2006). Models in science. In N. Z. Edward (Ed.). *The Stanford Encyclopedia of Philosophy (Spring 2006 Edition).* http://plato.stanford.edu/archives/spr2006/entries/models-science/

Gamma, E., Helm, R., Johnson, R., & Vlissides, J. (1994). *Design patterns.* Reading, MA: Addison-Wesley.

Gentner, D., & Stevens, A. L. (Eds.). (1983). *Mental models.* Hillsdale, NJ: Erlbaum.

Gibson, D., & Clarke-Midura, J. (2015). Some psychometric and design implications of game-based learning analytics. In P. Isaías, J. M. Spector, D. Ifenthaler, & D. G. Sampson (Eds.), *E-learning systems, environments and approaches* (pp. 247–261). Cham, Switzerland: Springer International Publishing.

Giere, R. N. (1987). The cognitive study of science. In N. J. Neressian (Ed.), *The process of science* (pp. 139–159). Dordrecht: Martinus Nijhoff.

Giere, R. N. (2004). How models are used to represent reality. *Philosophy of Science, 71,* 742–752.

Gilbert, J. K., & Justi, R. (2016). *Modelling-based teaching in science education.* Switzerland: Springer International Publishing.

Glaser, R., Chi, M. T., & Farr, M. J. (Eds.). (1988). *The nature of expertise.* Mahwah, NJ: Lawrence Erlbaum Associates.

Gobert, J., & Buckley, B. (2000). Special issue editorial: Introduction to model-based teaching and learning. *International Journal of Science Education, 22,* 891–894.

Gobert, J. D., Sao Pedro, M., Baker, R. S. J. D., Toto, E., & Montalvo, O. (2012). Leveraging educational data mining for real time performance assessment of scientific inquiry skills within microworlds. *Journal of Educational Data Mining, 5,* 153–185.

Gorin, J. S., & Mislevy, R. J. (2013). Inherent measurement challenges in the Next Generation Science Standards for both formative and summative assessment. Princeton, NJ: K-12 Center at ETS.

Gotwals, A. W., & Songer, N. B. (2010). Reasoning up and down a food chain: Using an assessment framework to investigate students' middle knowledge. *Science Education, 94,* 260–281.

Greeno, J. G. (1989). Situations, mental models, and generative knowledge. In D. Klahr & K. Kotovsky (Eds.), *Complex information processing* (pp. 285–318). Hillsdale, NJ: Lawrence Erlbaum.

Grosslight, L., Unger, C., Jay, E., & Smith, C. L. (1991). Understanding models and their use in science: Conceptions of middle and high school students and experts. *Journal of Research in Science Teaching, 28,* 799–822.

Haertel, G. D., Haydel DeBarger, A., Villalba, S., Hamel, L., & Mitman Colker, A. (2010). *Integration of evidence-centered design and universal design principles using PADI, an online assessment design system* (Assessment for Students with Disabilities Technical Report 3). Menlo Park, CA: SRI International. http://padi-se.sri.com/downloads/TR3_Integrating EDCandUDL.pdf

Hammer, D., Elby, A., Scherr, R. E., & Redish, E. F. (2005). Resources, framing, and transfer. In J. P. Mestre (Ed.), *Transfer of learning from a modern multidisciplinary perspective* (pp. 89–120). Greenwich, CT: Information Age Publishing.

Hansen, E. G., Mislevy, R. J., Steinberg, L. S., Lee, M. J., & Forer, D. C. (2005). Accessibility of tests within a validity framework. *System: An International Journal of Educational Technology and Applied Linguistics, 33,* 107–133.

Harris, C. J., Krajcik, J. S., Pellegrino, J. W., & McElhaney, K. W. (2016). *Constructing assessment tasks that blend disciplinary core Ideas, crosscutting concepts, and science practices for classroom formative applications.* Menlo Park, CA: SRI International.

Harrison, A. G., & Treagust, D. F. (2000). A typology of school science models. *International Journal of Science Education, 22,* 1011–1026.

Heller, P., & Heller, K. (2001). *Cooperative group problem solving in physics.* Pacific Grove, CA: Thomson Brooks/Cole.

Hesse, F., Care, E., Buder, J., Sassenberg, K., & Griffin, P. (2015). A framework for teachable collaborative problem solving skills. In P. Griffin & E. Care (Eds.), *Assessment and teaching of 21st century skills* (pp. 37–56). Dordrecht, The Netherlands: Springer.

Hestenes, D. (1987). Toward a modeling theory of physics instruction. *American Journal of Physics, 55,* 440–454.

Hestenes, D., Wells, M., & Swackhamer, G. (1992). Force concept inventory. *The Physics Teacher, 30,* 141–151.

Hunt, E., & Minstrell, J. (1994). A cognitive approach to the teaching of physics. In K. McGilly (Ed.), *Classroom lessons: Integrating cognitive theory and classroom practice* (pp. 51–74). Cambridge, MA: MIT Press.

Ingham, A. M., & Gilbert, J. K. (1991). The use of analogue models by students of chemistry at higher education level. *International Journal of Science Education, 13,* 193–202.

Johnson, S. K., & Stewart, J. (2002). Revising and assessing explanatory models in a high school class: A comparison of unsuccessful and successful performance. *Science Education, 86,* 463–480.

Johnson-Laird, P. N. (1983). Mental models: *Towards a cognitive science of language, inference, and consciousness.* Cambridge, MA: Harvard University Press.

Jungck, J. R., & Calley, J. (1985). Strategic simulations and post-Socratic pedagogy: Constructing computer software to develop long-term inference through experimental inquiry. *American Biology Teacher, 47,* 11–15.

Kahnemann, D. (2011). *Thinking, fast and slow.* NY: Farrar, Straus & Giroux.

Kalyuga, S. (2006). Rapid cognitive of learners' knowledge structures. *Learning and Instruction,* *16,* 1–11.

Kintsch, W. (1998). *Comprehension: A paradigm for cognition.* New York: Cambridge University Press.

Kintsch, W., & Greeno, J. G. (1985). Understanding and solving word arithmetic problems. *Psychological Review, 92,* 109–129.

Larkin, J. (1983). The role of problem representation in physics. In D. Gentner & Stevens, A. L. (Eds.). *Mental models* (pp. 75–98). Hillsdale, NJ: Lawrence Erlbaum Associates.

Latour, B. (1987). *Science in Action.* Cambridge, MA: Harvard University Press.

Lehrer, R., & Schauble, L. (2006). Cultivating in science education. In K. Sawyer (Ed.), *Cambridge handbook of the learning sciences* (pp. 371–388). New York: Cambridge University Press.

Linn, R. L. (2000). Assessments and accountability. *Educational Researcher, 29*(2), 4–16.

Liu, M., & Haertel, G. (2011). *Design patterns: A tool to support assessment task authoring (Large-scale Assessment Technical Report 11).* Menlo Park: SRI International.

Margolis, M. J., & Clauser, B. E. (2006). A regression-based procedure for automated scoring of a complex medical performance assessment. In D. M. Williamson, R. J. Mislevy, & I. I. Bejar (Eds.), *Automated scoring of complex in computer based testing* (pp. 132–167). Mahwah, NJ: Erlbaum.

Markman, A. B. (1999). *Knowledge representation.* Mahwah, NJ: Erlbaum.

Marshall, S. P. (1993). Assessing schema knowledge. In N. Frederiksen, R. J. Mislevy, & I. I. Bejar (Eds.), *Test theory for a new generation of tests* (pp. 155–180). Hillsdale, New Jersey: Lawrence Erlbaum.

Marshall, S. P. (1995). Schemas in problem solving. Cambridge University Press, New York.

Martin, J. D., & VanLehn, K. (1995). A Bayesian approach to cognitive. In P. Nichols, S. Chipman, & R. Brennan (Eds.), *Cognitively diagnostic assessment* (pp. 141–165). Hillsdale, NJ: Erlbaum.

Means, B. & Haertel, G. (2002). Technology supports for assessing science inquiry. In *Technology and assessment: Thinking ahead* (pp. 12–25). National Research Council. Washington, DC: National Academy Press.

Messick, S. J. (1989). Validity. In R. L. Linn (Ed.), *Educational measurement* (3rd ed., pp. 13–103). New York: Macmillan.

Messick, S. J. (1994). The interplay of evidence and consequences in the validation of performance assessments. *Educational Researcher, 23*(2), 13–23.

Mislevy, R. J. (2003). Substance and structure in assessment arguments. *Law, Probability, and Risk, 2,* 237–258.

Mislevy, R. J. (2006). Cognitive psychology and educational assessment. In R. L. Brennan (Ed.), *Educational measurement* (4th ed., pp. 257–305). Westport, CT: American Council on Education/Praeger.

Mislevy, R. J. (2017). Resolving the paradox of rich performance tasks. In H. Jiao & R. W. Lissitz (Eds.), *Test fairness in the new generation of large-scale assessment* (pp. 1–46). Charlotte, NC: Information Age Publishing.

Mislevy, R. J., Corrigan, S., Oranje, A., DiCerbo, K., John, M., Bauer, M. I., et al. (2014). *Psychometric considerations in game-based.* New York: Institute of Play.

Mislevy, R. J., Haertel, G., Cheng, B. H., Ructtinger, L., DeBarger, A., Murray, E., et al. (2013). A "conditional" sense of fairness in assessment. *Educational Research and Evaluation, 19,* 121–140.

Mislevy, R.J., Hamel, L., Fried, R., Gaffney, G., Haertel, T., Hafter G., et al. (2003). *Design patterns for assessing science inquiry (PADI Technical Report 1).* Menlo Park: SRI International.

Mislevy, R. J., & Riconscente, M. M. (2006). Evidence-centered assessment design: Layers, concepts, and terminology. In S. Downing & T. Haladyna (Eds.), *Handbook of test development* (pp. 61–90). Mahwah, NJ: Erlbaum.

Mislevy, R. J., Steinberg, L. S., & Almond, R. G. (2003). On the structure of educational assessments. *Measurement: Interdisciplinary Research and Perspectives, 1,* 3–67.

Mislevy, R. J., Steinberg, L. S., Breyer, F. J., Johnson, L., & Almond, R. A. (2002). Making sense of data from complex assessments. *Applied Measurement in Education, 15,* 363–378.

Mosteller, F., & Tukey, J. W. (1977). *Data analysis and regression: A second course in statistics.* Reading, MA: Addison-Wesley.

National Research Council. (1996). *National Science Education Standards.* Washington, DC: National Academy Press.

National Research Council (2000). How people learn: Brain, mind, experience and school (2nd ed.). In J. D. Bransford, A. L. Brown, & R. Cocking (Eds.). *Committee on developments in the science of learning.* Washington, DC: National Academy Press.

National Research Council. (2001). *Knowing what students know: The science and design of educational.* Committee on the Foundations of Assessment, J. Pellegrino, R. Glaser, & N. Chudowsky (Eds.), Washington DC: National Academy Press.

National Research Council. (2012). *A framework for K-12 science education: Practices, crosscutting concepts, and core ideas.* Committee on a Conceptual Framework for New K-12 Science Education Standards. Board on Science Education, Division of Behavioral and Social Sciences and Education. Washington, DC: The National Academies Press.

Newell, A., & Simon, H.A. (1972). *Human problem solving.* Englewood Cliffs, NJ: Prentice-Hall.

NGSS Lead States. (2013a). *How to read the Next Generation Science Standards.* Retrieved January 23, 2016 from http://www.nextgenscience.org/sites/ngss/files/How%20to%20Read%20NGSS%20-%20Final%204-19-13.pdf

NGSS Lead States. (2013b). *Next Generation Science Standards: For States, By States.* Washington, DC: The National Academies Press.

Norman, D. A. (1993). *Things that make us smart.* Boston: Addison-Wesley.

Novick, S., & Nussbaum, J. (1981). Pupils' understanding of the particulate nature of matter: A cross-age study. *Science Education, 65,* 187–196.

Pellegrino, J. W. (2013). Proficiency in science: Assessment challenges and opportunities. *Science, 340,* 320–323.

Perkins, D. N., & Salomon, G. (1989). Are cognitive skills context-bound? *Educational Researcher, 18,* 16–25.

Piaget, J. (1976). Piaget's theory. In B. Inhelder, H. H. Chipman, & C. Zwingmann (eds.). *Piaget and his school* (pp. 11–23). Springer: Berlin Heidelberg.

Quellmalz, E. S., Timms, M. J., Buckley, B. C., Davenport, J., Loveland, M., & Silberglitt, M. D. (2012). 21st century dynamic assessment. In M. Mayrath, J. Clarke-Midura, & D. H. Robinson (Eds.), *Technology-based assessments for 21st century skills: Theoretical and practical implications from modern research.* Charlotte, NC: Information Age.

Redish, E. F. (2004). A theoretical framework for physics education research: Modeling student thinking. In E. F. Redish & M. Vicentini (Eds.), *Proceedings of the Enrico Fermi Summer School Course, CLVI* (pp. 1–63). Bologna, Italy: Italian Physical Society.

Richmond, B. (2005). *An introduction to systems thinking, featuring STELLA.* Lebanon, NH: isee systems.

Riconscente, M. M., Mislevy, R. J., & Corrigan, S. (2015). In S. Lane, T. M. Haladyna, & M. Raymond (Eds.), *Handbook of test development* (2nd ed., pp. 40–63). Routledge: Informa, Taylor & Francis.

Riley, M. S., Greeno, J. G., & Heller, J. I. (1983). Development of children's problem-solving ability in arithmetic. In H. P. Ginsburg (Ed.), *The development of mathematical thinking* (pp. 153–196). New York: Academic Press.

Rose, D, Murray, E., & Gravel, J. (2012). UDL and the PADI process: The foundation (Assessment for Students with Disabilities Technical Report 4). Menlo Park, CA: SRI International. http://padi-se.sri.com/downloads/TR4_UDL_Tech_Report2012FL.pdf

Ruiz-Primo, M. A., & Shavelson, R. J. (1996). Rhetoric and reality in science performance assessments: An update. *Journal of Research in Science Teaching, 33,* 1045–1063.

Rumelhart, D. E., & Norman D. A. (1977). Accretion, tuning and restructuring: Three modes of learning. In J. W. Cotton & R. L. Klatzky (Eds.), *Semantic factors in cognition* (pp. 37–54). Hillsdale, NJ: Erlbaum.

Rupp, A. A., Levy, R., DiCerbo, K. E., Sweet, S., Crawford, A. V., Caliço, T., et al. (2012). Putting ECD into practice: The interplay of theory and data in evidence models within a digital learning environment. *Journal of Educational Data Mining, 4,* 49–110.

Scalise, K., & Gifford, B. (2006). Computer-based in E-Learning: A framework for constructing "Intermediate Constraint" questions and for technology platforms. *Journal of Technology, Learning, and Assessment, 4*(6). Retrieved July 17, 2009, from http://www.jtla.org

Schunn, C. D., & Anderson, J. R. (1999). The generality/specificity of expertise in scientific reasoning. *Cognitive Science, 23,* 337–370.

Schwarz, C. V., Reiser, B. J., Davis, E. A., Kenyon, L., Achér, A., Fortus, D., … & Krajcik, J. (2009). Developing a learning progression for scientific modeling: Making scientific modeling accessible and meaningful for learners. *Journal of Research in Science Teaching, 46* (6), 632–654.

Seibert, G., Hamel, L., Haynie, K., Mislevy, R., & Bao, H. (2006). *Mystery powders: An application of the PADI Design System using the Four-Process Delivery System (PADI Technical Report 15).* Menlo Park, CA: SRI International.

Shute, V. J., Masduki, I., Donmez, O., Dennen, V. P., Kim, Y. J., Jeong, A. C., Wang, C. (2010). Modeling, assessing, and supporting key competencies within game environments. In D. Ifenthaler, P. Pirnay-Dummer, & N. M. Seel (Eds.), *Computer-based diagnostics and systematic analysis of knowledge* (pp. 281–309). New York: Springer.

Shute, V. J., Ventura, M., Bauer, M., & Zapata-Rivera, D. (2009). Melding the power of serious games and embedded to monitor and foster learning. In U. Ritterfeld, M. Cody, & P. Vorderer (Eds.), *Serious games: Mechanisms and effects* (pp. 295–321). New York: Routledge.

Simon, H. A. (1996). *The sciences of the artificial.* Cambridge, MA: MIT Press.

Snir, J., Smith, C. L., & Raz, G. (2003). Linking with competing underlying models: A software tool for introducing students to the particulate nature of matter. *Science Education, 87,* 794–830.

Songer, N. B., Kelcey, B., & Gotwals, A. W. (2009). How and when does complex reasoning occur? Empirically driven development of a learning progression focused on complex reasoning about biodiversity. *Journal of Research in Science Teaching, 46,* 610–631.

Spitulnik, M. W., Krajcik, J., & Soloway, E. (1999). Construction of models to promote scientific understanding. In W. Feurzeig & N. Roberts (Eds.), *Modeling and simulation in science and mathematics education* (pp. 70–94). New York: Springer-Verlag.

Steinberg, L. S., & Gitomer, D. G. (1996). Intelligent tutoring and built on an understanding of a technical problem-solving task. *Instructional Science, 24,* 223–258.

Steinberg, L. S., Mislevy, R. J., Almond, R. G., Baird, A. B., Cahallan, C., Dibello, L. V., … Kindfield, A. C. (2003). *Introduction to the biomass project: An illustration of evidence-centered assessment design and delivery capability (CSE Technical Report 609).* Los Angeles: The National Center for Research on Evaluation, Standards, Student Testing (CRESST), Center for Studies in Education, UCLA.

Stewart, J., & Hafner, R. (1991). Extending the conception of "problem" in problem-solving research. *Science Education, 75*(1), 105–120.

Stewart, J., & Hafner, R. (1994). Research on problem solving: Genetics. In D. Gabel (Ed.), *Handbook of research on science teaching and learning* (pp. 284–300). New York: MacMillan.

Stewart, J., Hafner, R., Johnson, S., & Finkel, E. (1992). Science as model-building: Computers and high school genetics. *Educational Psychologist, 27*(3), 317–336.

Stewart, J., Passmore, C., Cartier, J., Rudolph, J., & Donovan, S. (2005). Modeling for understanding science education. In T. Romberg, T. Carpenter, & F. Dremock (Eds.), *Understanding mathematics and science matters* (pp. 159–184). Mahwah, N.J.: Lawrence Erlbaum.

Strauss, C., & Quinn, N. (1998). *A cognitive theory of cultural meaning*. New York: Cambridge University Press.

Suárez, M. (2004). An inferential conception of scientific representation. *Philosophy of Science, 71*, 767–779.

Swoyer, C. (1991). Structural representation and surrogative reasoning. *Synthese, 87*, 449–508.

Thornton, R. K., & Sokoloff, D. R. (1998) Assessing student learning of Newton's laws: The force and motion conceptual evaluation and the evaluation of active learning laboratory and lecture curricula. *American Journal of Physics, 66*, 338–351.

Toulmin, S. E. (1958). *The uses of argument*. Cambridge, England: Cambridge University Press.

Wertsch, J. V. (1998). *Mind as action*. Oxford: Oxford University Press.

West, P., Wise Rutstein, D., Mislevy, R. J., Liu, J., Levy, R., DiCerbo, K. E., et al. (2012). A Bayesian network approach to modeling learning progressions. In A. C. Alonzo & A. W. Gotwals (Eds.), *Learning progressions in science* (pp. 255–291). Rotterdam: Sense Publishers.

White, B. Y., & Frederiksen, J. R. (1998). Inquiry, modeling, and metacognition: Making science accessible to all students. *Cognition and Instruction, 16*, 3–118.

White, B. Y., Shimoda, T. A., & Frederiksen, J. R. (1999). Enabling students to construct theories of collaborative inquiry and reflective learning: Computer support for metacognitive development. *International Journal of Artificial Intelligence in Education, 10*, 151–182.

Wiley, D. E., & Haertel, E. H. (1996). Extended assessment tasks: Purposes, definitions, scoring, and accuracy. In M. B. Kane & R. Mitchell (Eds.), *Implementing performance assessments: Promises, problems, and challenges*. Mahwah, NJ: Erlbaum.

Zalles, D., Haertel, G., & Mislevy, R. (2010). *Using evidence-centered design to support assessment, design and validation of learning progressions* (Large-Scale Assessment Technical Report 10). Menlo Park, CA: SRI International. http://ecd.sri.com/downloads/ECD_TR10_Learning_Progressions.pdf

Index

© The Author(s) 2017
R.J. Mislevy et al., *Assessing Model-Based Reasoning using Evidence-Centered Design*, SpringerBriefs in Statistics, DOI 10.1007/978-3-319-52246-3

CPSIA information can be obtained
at www.ICGtesting.com
Printed in the USA
BVOW07s0800140817

492007BV00003B/7/P